T0142012

Studies in Computational Intelligence

Volume 561

Series editor

Janusz Kacprzyk, Polish Academy of Sciences, Warsaw, Poland
e-mail: kacprzyk@ibspan.waw.pl

For further volumes:
http://www.springer.com/series/7092

About this Series

The series "Studies in Computational Intelligence" (SCI) publishes new developments and advances in the various areas of computational intelligence—quickly and with a high quality. The intent is to cover the theory, applications, and design methods of computational intelligence, as embedded in the fields of engineering, computer science, physics and life sciences, as well as the methodologies behind them. The series contains monographs, lecture notes and edited volumes in computational intelligence spanning the areas of neural networks, connectionist systems, genetic algorithms, evolutionary computation, artificial intelligence, cellular automata, self-organizing systems, soft computing, fuzzy systems, and hybrid intelligent systems. Of particular value to both the contributors and the readership are the short publication timeframe and the world-wide distribution, which enable both wide and rapid dissemination of research output.

Valentina Emilia Balas
Petia Koprinkova-Hristova
Lakhmi C. Jain
Editors

Innovations in Intelligent Machines-5

Computational Intelligence in Control Systems Engineering

 Springer

Editors
Valentina Emilia Balas
Department of Automation and Applied
  Informatics
"Aurel Vlaicu" University of Arad
Arad
Romania

Lakhmi C. Jain
Faculty of Education, Science, Technology
  and Mathematics
University of Canberra
Canberra
Australia

Petia Koprinkova-Hristova
Institute of Information and Communication
  Technologies
Bulgarian Academy of Sciences
Sofia
Bulgaria

ISSN 1860-949X   ISSN 1860-9503   (electronic)
ISBN 978-3-662-52156-4   ISBN 978-3-662-43370-6   (eBook)
DOI 10.1007/978-3-662-43370-6
Springer Heidelberg New York Dordrecht London

Printed on acid-free paper

Springer is part of Springer Science+Business Media (www.springer.com)

Foreword

It is desired to design machines which assist people in their everyday lives. Many researchers in related research areas have been dreaming of such intelligent machines which can be used under various situations. I am pleased to find a series of volumes in the field of Intelligent Machines published by Springer:

- Chahl, J.S., Jain, L.C., Mizutani, A. and Sato-Ilic, M., Innovations in Intelligent Machines 1, Springer-Verlag, Germany, 2007.
- Watanabe, T. and Jain, L.C., Innovations in Intelligent Machines 2: Intelligent Paradigms and Applications, Springer-Verlag, Germany, 2012.
- Jordanov, I. and Jain, L.C., Innovations in Intelligent Machines 3: Contemporary Achievements in Intelligent Systems, Springer-Verlag, Germany, 2012.
- Faucher, C. and Jain, L.C., Innovations in Intelligent Machines 4: Recent Advances in Knowledge Engineering, Springer-Verlag, Germany, 2014.

A new volume *Innovations in Intelligent Machines-5* deals with recent advances in computational intelligence on control systems engineering. Professor Valentina Balas and her Co-Editors present excellent applications to demonstrate the advantages of intelligence-based systems. I have known Prof. Balas for many years and been impressed by her dedication to the service of engineering and science profession such as undertaking excellent research projects, editing books and an international journal, presenting keynote addresses, and so on.

I am confident that this research volume will arouse great interest in the scientific community and will encourage readers to further continue their research in this field.

Japan, May 2014   Michio Sugeno

Preface

The research book includes some of the most recent research in the theoretical foundation and practical applications of control systems engineering. The chapters in this book present an overview of various applications for developing advanced computational intelligent methods in control systems in different areas. Due to the variety and complexity of the systems which involve vagueness, imprecision, and uncertainty, neural and fuzzy systems represent suitable approaches for control systems. They offer many advantages over conventional control methods.

The book includes eight chapters and presents various computational paradigms in control systems engineering. A number of applications and case study are also introduced.

Chapter 1 by Ieroham Baruch and Eloy Echeverria Saldierna present the use of a Recurrent Neural Network Model (RNNM) incorporated in a fuzzy-neural multimodel for decentralized identification of an aerobic digestion process. The analytical model of the digestion bioprocess represented a distributed parameter system, which is reduced to a lumped system using the orthogonal collocation method, applied in four collocation points. The proposed decentralized RNNM consists of five independently working Recurrent Neural Networks (RNN), so to approximate the process dynamics in four different measurement points plus the recirculation tank.

Chapter 2 by Petia Georgieva and Sebastião Feyo de Azevedo is focused on developing of a feasible model predictive control (MPC) based on time-dependent recurrent neural network (NN) models. A modification of the classical regression neural models is proposed suitable for prediction purposes. In order to reduce the computational complexity and to improve the prediction ability of the neural model, optimization of the NN structure (lag space selection, number of hidden nodes), pruning techniques, and identification strategies are discussed. Furthermore, a computationally efficient modification of the general nonlinear MPC is proposed termed Error Tolerant MPC (ETMPC). The NN model is imbedded into the structure of the ETMPC and extensively tested on a dynamic simulator of an industrial crystallizer.

In Chap. 3 by Nikolaos A. Sofianos and Yiannis S. Boutalis, the recent developments in the field of Intelligent Multiple Models-based Adaptive Switching Control (IMMASC) are discussed. It provides at the same time all the essential information about the conventional single model and multiple models adaptive control, which constitute the base for the development of the new intelligent methods. The work emphasizes on the importance and the advantages of IMMASC in the field of control systems technology presenting control structures that contain linear robust models, neural models, and T-S fuzzy models. One of the main advantages of switching control systems against the single model control architectures is that they are able to provide stability and improved performance in multiple environments when the systems to be controlled have unknown parameters or highly uncertain parameters. Some hybrid multiple models control architectures are presented, and a numerical example is given in order to illustrate the efficiency of the intelligent methods.

Chapter 4 by D. Vijay Rao and V. V. S. Sharma presents the integration of case-based reasoning and decision theory based on Computational Intelligence techniques and its usefulness in the retrieval and selection of reusable software components from a software components repository. Software components are denoted by cases with a set of features, attributes, and relations of a given situation and its associated outcomes. These are taken as inputs to a Decision Support tool that classifies the components as adaptable to the given situation with membership values for the decisions. This classification is based on Rough-Fuzzy set theory and the methodology is explained with illustrations. In this novel approach, CBR and DSS (based on Rough-Fuzzy sets) have been applied successfully to the software engineering domain to address the problem of retrieving suitable components for reuse from the case data repository. A software tool called RuFTool is developed as a decision support tool for component retrieval for reuse on Windows platform with MS-ACCESS as the back-end and Visual BASIC as the front-end to this purpose. The use of rough-fuzzy sets increases the likelihood of finding the suitable components for reuse when exact matches are not available or are very few in number.

Chapter 5 by D. Vijay Rao and Dana Balas-Timar is stressing on considerable efforts on designing military training simulators using modeling, simulation, and analysis for operational analyses and training. Air Warfare Simulation System is an agent-oriented virtual warfare simulator that is designed using these concepts for operational analysis and course of action analysis for training. A critical factor that decides the next course of action and hence the results of the simulation is the skill, experience, situation awareness of the pilot in the aircraft cockpit and the pilots' decision-making ability in the cockpit. Advances in combat aircraft avionics and on-board automation, information from on-board and ground sensors, and satellites poses a threat in terms of information and cognitive overload to the pilot, and triggering conditions that makes decision making a difficult task. The authors describe a novel approach based on soft computing and computational intelligence paradigms called ANFIS, a neuro-fuzzy hybridization technique, to model the pilot agent and its behavior characteristics in the warfare simulator. This emerges as an interesting problem as the decisions made are dynamic and depend upon the actions taken by enemy. It is also build a pilots' database that represents the specific cognitive characteristics, skills, training experience. Authors illustrate the methodology with suitable examples and lessons drawn from the virtual air warfare simulator.

Chapter 6 by Juš Kocijan and Alexandra Grancharova explain that systems can be characterized as complex since they have a nonlinear behavior incorporating a stochastic uncertainty. They show that one of the most appropriate methods for modeling of such systems is based on the application of Gaussian processes (GPs). The GP models provide a probabilistic nonparametric modeling approach for black-box identification of nonlinear stochastic systems. This chapter reviews the methods for modeling and control of complex stochastic systems based on GP models. The GP-based modeling method is applied in a process engineering case study, which represents the dynamic modeling and control of a laboratory gas–liquid separator. The variables to be controlled are the pressure and the liquid level in the separator and the manipulated variables are the apertures of the valves for the gas flow and the liquid flow. GP models with different regressors and different covariance functions are obtained and evaluated. A selected GP model of the gas–liquid separator is further used to design an explicit stochastic model predictive controller to ensure the optimal control of the separator.

In Chap. 7 by N. Paraschiv, M. Oprea, M. Cărbureanu, and M. Olteanu are introduced two computational intelligence techniques, genetic algorithms and neuro-fuzzy systems, for chemical process control. The authors present the objectives and conventional automatic control of chemical processes and the computational intelligence techniques for process control. A case study that describes a neuro-fuzzy control system for a wastewater pH neutralization process is presented in detail.

Chapter 8 by R. Krishna Priya, C. Thangaraj, C. Kesavadas, and S. Kannan is focused on image segmentation technique based on Modified Particle Swarm optimized—fuzzy entropy applied for Infra Red (IR) images to detect the object of interest and Magnetic Resonance (MR) brain images to detect a brain tumor. Adaptive thresholding of input IR images and MR images are performed based on the proposed method. The input image is classified into dark and bright parts with Membership Functions (MF), whose member functions of the fuzzy region are Z-function and S-function. The optimal combination of parameters of these fuzzy MFs are obtained using Modified Particle Swarm Optimization (MPSO) algorithm. The objective function for obtaining the optimal fuzzy MF parameters is considered to be the maximum the fuzzy entropy. Through numerous examples, the performance of the proposed method is compared with those using existing entropy-based object segmentation approaches and the superiority of the proposed method is demonstrated. The experimental results obtained are compared with the enumerative search method and Otsu segmentation technique.

We believe that scientists, engineers, professors, students, and all interested in this subject will find the book useful and interesting.

The book is due to the excellent contribution by the authors. We are grateful to reviewers for their constructive and enlightening comments. The excellent editorial assistance by the Springer is acknowledged for the excellent collaboration and patience during the evolvement of this volume.

Romania, March 2014   Valentina Emilia Balas
Bulgaria   Petia Koprinkova-Hristova
Australia   Lakhmi C. Jain

Contents

Chapter 1
Decentralized Fuzzy-Neural Identification and I-Term Adaptive Control of Distributed Parameter Bioprocess Plant

Ieroham Baruch and Eloy Echeverria Saldierna

Abstract The chapter proposed to use of a Recurrent Neural Network Model (RNNM) incorporated in a fuzzy-neural multi model for decentralized identification of an aerobic digestion process, carried out in a fixed bed and a recirculation tank anaerobic wastewater treatment system. The analytical model of the digestion bioprocess represented a distributed parameter system, which is reduced to a lumped system using the orthogonal collocation method, applied in four collocation points. The proposed decentralized RNNM consists of five independently working Recurrent Neural Networks (RNN), so to approximate the process dynamics in four different measurement points plus the recirculation tank. The RNN learning algorithm is the second order Levenberg-Marquardt one. The comparative graphical simulation results of the digestion wastewater treatment system approximation, obtained via decentralized RNN learning, exhibited a good convergence, and precise plant variables tracking. The identification results are used for I-term direct and indirect (sliding mode) control obtaining good results.

1.1 Introduction

In the last decade, the Computational Intelligence tools (CI), like Artificial Neural Networks (ANN), Fuzzy Systems (FS), and its hybrid neuro-fuzzy and fuzzy-neural systems, became universal means for many applications in identification, prediction and control. Because of their approximation and learning capabilities,

I. Baruch (✉) · E. E. Saldierna
Department of Automatic Control, CINVESTAV-IPN, Ave. IPN No 2508, A.P. 14-470
07360 Mexico, D.F., Mexico
e-mail: baruch@ctrl.cinvestav.mx

E. E. Saldierna
e-mail: eecheverria@ctrl.cinvestav.mx

V. E. Balas et al. (eds.), *Innovations in Intelligent Machines-5*,
Studies in Computational Intelligence 561, DOI: 10.1007/978-3-662-43370-6_1,
© Springer-Verlag Berlin Heidelberg 2014

the ANNs have been widely employed to dynamic process modeling and control, including biotechnological plants, [1–9]. Among several possible neural network architectures the ones most widely used are the Feedforward NN (FFNN) and the Recurrent NN (RNN), [1, 2]. The main NN property namely the ability to approximate complex non-linear relationships without prior knowledge of the model structure makes them a very attractive alternative to the classical modeling and control techniques [2]. Also, a great boost has been made in the applied NN-based adaptive control methodology incorporating integral plus state control action in the control law, [10]. The FFNN and the RNN have been applied for Distributed Parameter Systems (DPS) identification and control too [11–17]. In [11], a RNN is used for system identification and process prediction of a DPS dynamics—an adsorption column for wastewater treatment of water contaminated with toxic chemicals. Similarly to the static ANNs, the fuzzy models could approximate static nonlinear plants where structural plants information is needed to extract the fuzzy rules, [18–27]. The aim of the fuzzy-neural models is to merge both ANN and FS approaches so to obtain fast adaptive models possessing learning, [18]. The fuzzy-neural networks are capable to incorporate both numerical data (quantitative information), and expert's knowledge (qualitative information) and describe them in the form of linguistic IF-THEN rules. During the last decade considerable research has been devoted towards developing recurrent neuro-fuzzy (fuzzy-neural) models, summarized in [20]. To reduce the number of IF-THAN rules, the hierarchical approach could be used [20]. A promising approach of recurrent fuzzy-neural systems with internal dynamics is the application of the Takagi-Sugeno (T-S) fuzzy rules with a static premise and a dynamic functional consequent part, [20]. The paper of [20] proposed as a dynamic function in the consequent part of the T-S rules to use a Recurrent Neural Network Model (RNNM). Together with the Recurrent Trainable Neural Network (RTNN) topology, the Backpropagation (BP) learning algorithm [20] is incorporated in the learning procedure taking part in the IF-THAN T-S rule antecedent. To complete the fuzzy-neural system learning, a second hierarchical defuzzification BP learning level has been formed so to improve the adaptation ability of the system, [20]. The aim of this chapter is to describe the results obtained by this system for decentralized identification and control of wastewater treatment anaerobic digestion bioprocess [25–28], representing a DPS, extending the used direct and indirect (sliding mode-SM) control laws with integral terms, so to form integral plus state control actions, capable to speed up the reaction of the control system and to augment its resistance to process and measurement noises, [11, 12]. The analytical anaerobic bioprocess plant model [25–28], used as an input/output plant data generator, is described by Partial Differential Equations (PDE)/Ordinary Differential Equations (ODE), and simplified using the Orthogonal Collocation Method (OCM), [29], in four collocation points for the fixed bed, plus one-in a recirculation tank. These measurement points are used as centres of the membership functions of the fuzzified space variables of the plant. The learning algorithm used in the antecedent part of the

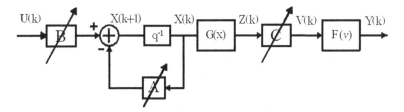

Fig. 1.1 Block diagram of the RTNN model

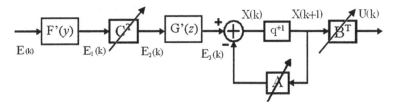

Fig. 1.2 Block diagram of the adjoint RTNN model

T-S identification rules is the Levenberg-Marquardt's learning which is faster than the BP one, [30]. For sake of clarity all abbreviations used in this chapter are summarized in a Table A.1 and given in Appendix 2.

1.2 Description of the RTNN Topology and Learning

This part described the RTNN topology and both the backpropagation first order learning and the Levenberg-Marquardt second order learning of this RTNN.

1.2.1 Description of the RTNN Topology and Its Real-Time BP Learning

Block-diagrams of the RTNN topology and its adjoint, are given on Figs. 1.1 and 1.2.

Following Figs. 1.1 and 1.2, we could derive the dynamic BP algorithm of its learning based on the RTNN topology using the diagrammatic method of [31]. The RTNN topology and learning are described in vector-matrix form as:

$$X(k+1) = AX(k) + BU(k); \quad B = [B_1; B_0]; \quad U^T = [U_1; U_2]; \quad (1.1)$$

$$Z_1(k) = G[X(k)]; \quad (1.2)$$

$$V(k) = CZ(k); \; C = [C_1 \mid C_0]; \; Z^T = [Z_1 \mid Z_2]; \tag{1.3}$$

$$Y(k) = F[V(k)]; \tag{1.4}$$

$$A = block-diag\,(Ai), \; |Ai| < 1; \tag{1.5}$$

$$W(k+1) = W(k) + \eta\,\Delta W(k) + \alpha\,\Delta W(k-1); \tag{1.6}$$

$$E(k) = T(k) - Y(k); \tag{1.7}$$

$$E_1(k) = F'[Y(k)]E(k); \; F'[Y(k)] = \left[1 - Y^2(k)\right]; \tag{1.8}$$

$$\Delta C(k) = E_1(k)Z^T(k); \tag{1.9}$$

$$E_3(k) = G'[Z(k)]\,E_2(k); \; E_2(k) = C^T(k)\,E_1(k); \; G'[Z(k)] = \left[1 - Z^2(k)\right]; \tag{1.10}$$

$$\Delta B(k) = E_3(k)\,U^T(k); \tag{1.11}$$

$$\Delta A(k) = E_3(k)\,X^T(k); \tag{1.12}$$

$$Vec(\Delta A(k)) = E_3(k) \cdot X(k); \tag{1.13}$$

where: X, Y, U are state, augmented output, and input vectors with dimensions N, $(L + 1)$, $(M + 1)$, respectively, where Z_1 and U_1 are the $(N \times 1)$ output and $(M \times 1)$ input of the hidden layer; the constant scalar threshold entries are $Z_2 = -1$, $U_2 = -1$, respectively; V is a $(L \times 1)$ pre-synaptic activity of the output layer; T is the $(L \times 1)$ plant output vector, considered as a RNN reference; A is $(N \times N)$ block-diagonal weight matrix; B and C are $[N \times (M + 1)]$ and $[L \times (N + 1)]$—augmented weight matrices; B_0 and C_0 are $(N \times 1)$ and $(L \times 1)$ threshold weights of the hidden and output layers; F[·], G[·] are vector-valued tanh(·)-activation functions with corresponding dimensions; F'[·], G'[·] are the derivatives of these tanh(·) functions; W is a general weight, denoting each weight matrix (C, A, B) in the RTNN model, to be updated; ΔW (ΔC, ΔA, ΔB), is the weight correction of W; η, α are learning rate parameters; ΔC is an weight correction of the learned matrix C; ΔB is an weight correction of the learned matrix B; ΔA is an weight correction of the learned matrix A; the diagonal of the matrix A is denoted by Vec(·) and Eq. (1.13) represents its learning as an element-by-element vector products; E, E_1, E_2, E_3, are error vectors with appropriate dimensions, predicted by the adjoint RTNN model, given on Fig. 1.2. The stability of the RTNN model is assured by the activation functions $(-1, 1)$ bounds and by the local stability weight bound condition, given by (1.39). Below a theorem of RTNN stability which represented an extended version of Nava's theorem, [32, 33], is given.

Theorem of Stability of the RTNN

Let the RTNN with Jordan Canonical Structure is given by Eqs. (1.1)–(1.5) (see Fig. 1.1) and the nonlinear plant model, is as follows:

$$X_d.(k + 1) = G[X_d(k), U(k)]$$
$$Y_d(k) = F[X_d(k)]$$

where: $\{Y_d(\cdot), X_d(\cdot), U(\cdot)\}$ *are output, state and input variables with dimensions* l, n_d, m, *respectively;* $F(\cdot)$, $G(\cdot)$ *are vector valued nonlinear functions with respective dimensions. Under the assumption of RTNN identifiability made, the application of the BP learning algorithm for* $A(\cdot)$, $B(\cdot)$, $C(\cdot)$, *in general matricial form, described by Eqs. (1.6)–(1.13), and the learning rates* $\eta(k)$, $\alpha(k)$ *(here they are considered as time-dependent and normalized with respect to the error) are derived using the following Lyapunov function:*

$$L(k) = L_1(k) + L_2(k)$$

where: $L_1(k)$ *and* $L_2(k)$ *are given by:*

$$L_1(k) = \frac{1}{2}e^2(k)$$

$$L_2(k) = tr\left(\tilde{W}_A(k)\tilde{W}_A^T(k)\right) + tr\left(\tilde{W}_B(k)\tilde{W}_B^T(k)\right) + tr\left(\tilde{W}_C(k)\tilde{W}_C^T(k)\right)$$

where:

$$\tilde{W}_A(k)=\widehat{A}(k) - A^*, \tilde{W}_B(k)=\widehat{B}(k) - B^*, \tilde{W}_C(k)=\widehat{C}(k) - C^*$$

are vectors of the estimation error and (A^*, B^*, C^*), $(\widehat{A}(k),\widehat{B}(k),\widehat{C}(k))$ *denote the ideal neural weight and the estimate of the neural weight at the kth step, respectively, for each case. Then the identification error is bounded,* i.e.:

$$L(k + 1)=L_1(k + 1) + L_2(k + 1)<0$$
$$\Delta L(k + 1) = L(k + 1) - L(k)$$

where the condition for $L_1(k + 1) < 0$ *is that:*

$$\frac{\left(1-\frac{1}{\sqrt{2}}\right)}{\psi_{max}} < \eta_{max} < \frac{\left(1+\frac{1}{\sqrt{2}}\right)}{\psi_{max}}$$

and for $L_2(k + 1) < 0$ we have:

$$\Delta L_2(k + 1) < -\eta_{max}|e(k + 1)|^2 - \alpha_{max}|e(k)|^2 + d(k + 1)$$

Note that η_{max} changes adaptively during the RTNN learning and:

$$\eta_{max} = \max_{i=1}^{3}\{\eta_i\}$$

where all: the unmodelled dynamics, the approximation errors and the pertur-bations, are represented by the d-term. Applying the Lemma of RTNN convergence for the given above result for $L_2(k + 1) < 0$ we could conclude that: the d-term must be bounded by the weight matrices and the learning parameter, in order to obtain the final result:

$$\Delta L_2(k) \in L_\infty$$

As a consequence we obtained $A_{(k)} \in L_\infty$, $B_{(k)} \in L_\infty$, $C_{(k)} \in L_\infty$. The complete proof of the theorem of stability is given in [32, 33].

Lemma of RTNN convergence
Applying the limit's definition, the identification error bound condition is obtained as:

$$\overline{\lim_{t \to \infty}} \frac{1}{k} \sum_{t=1}^{k} \left(|e_t|^2 + |e_{t-1}|^2 \right) \le d$$

Proof Starting from the final result of theorem of RTNN stability:

$$\Delta L_2(k) \le -\eta_{max}|e(k)|^2 - \alpha_{max}|e(k - 1)|^2 + d$$

After the analysis done, we get:

$$L_2(k + 1) - V(0) \le -\sum_{t=1}^{k} |e_t|^2 - \sum_{t=1}^{k} |e_{t-1}|^2 + d * k$$

$$\sum_{t=1}^{k} \left(|e_t|^2 + |e_{t-1}|^2 \right) \le d * k - L_2(k + 1) + L_2(0) \le d * k + L_2(0)$$

After this, let us divide by k, and applying the limit's definition, the identifi-cation error bound condition is obtained in the final form:

$$\overline{\lim_{t \to \infty}} \frac{1}{k} \sum_{t=1}^{k} \left(|e_t|^2 + |e_{t-1}|^2 \right) \le d$$

From here we can see that the term d must be bounded by weight matrices and the learning parameter, in order to obtain:

$$\Delta L_2(k) \in L_\infty$$

1.2.2 Description of the Real-Time Second-Order Levenberg-Marquardt Learning

The Levenberg-Marquardt (L-M) recursive algorithm of learning, [30], could be considered as a continuation of the BP algorithm and it will be used here. Following Fig. 1.1, the RTNN topology could be described in vector-matrix form as it is given by the Eqs. (1.1)–(1.5). The general recursive L-M algorithm of learning, [30], is given by the following equations:

$$W(k+1) = W(k) + P(k)DY[W(k)]^T E[W(k)] \qquad (1.14)$$

$$Y[W(k)] = g[W(k), U(k)] \qquad (1.15)$$

$$E^2[W(k)] = \left\{ Y_p(k) - g[W(k), U(k)] \right\}^2 \qquad (1.16)$$

$$DY[W(k)] = \left. \frac{\partial g[W(k), U(k)]}{\partial W} \right|_{W=W(k)} \qquad (1.17)$$

Where: W is a general weight matrix (A, B, C) under modification; P is the covariance matrix of the estimated weights updated; DY [·] is a Jacobean matrix with dimension $L \times N_w$; Y is the RTNN output vector which depends of the updated weights and the input; E is an error vector; Yp is the plant output vector, which is in fact the target vector. Using the same RTNN adjoint block diagram (see Fig. 1.2), it was possible to obtain the values of the matrix DY[·] for each updated weight, propagating the value D(k) = I through it. Applying Eq. (1.17) for each element of the weight matrices (A, B, C) in order to be updated, the corresponding gradient components are obtained as:

$$DY\big[C_{ij}(k)\big] = D_{1,i}(k)Z_j(k), \ D_{1,i}(k) = F_j'[Y_i(k)] \qquad (1.18)$$

$$DY\big[A_{ij}(k)\big] = D_{2,i}(k)X_j(k), \qquad (1.19)$$

$$DY\big[B_{ij}(k)\big] = D_{2,i}(k)U_j(k), \qquad (1.20)$$

$$D_{2,i}(k) = G'_i\big[Z_j(k)\big]C_iD_{1,i}(k).\tag{1.21}$$

Therefore the Jacobean matrix could be formed as:

$$DY[W(k)] = \big[DY\big(C_{ij}(k)\big), DY\big(A_{ij}(k)\big), DY\big(B_{ij}(k)\big)\big].\tag{1.22}$$

The $P(k)$ matrix was computed recursively by the equation:

$$P(k) = \alpha^{-1}(k)\big\{P(k-1) - P(k-1)\Omega^T[W(k)]S^{-1}[W(k)]\Omega[W(k)]P(k-1)\big\}\tag{1.23}$$

Where the $S(\cdot)$, and $\Omega(\cdot)$ matrices were given as follows:

$$S[W(k)] = \alpha(k)\Lambda(k) + \Omega[W(k)]P(k-1)\Omega^T[W(k)],\tag{1.24}$$

$$\Omega[W(k)] = \begin{bmatrix} 0 & \cdots & \begin{matrix} DY[W(k)] \\ 1 \end{matrix} & \cdots & 0 \end{bmatrix};\tag{1.25}$$

$$\Lambda(k)^{-1} = \begin{bmatrix} I & 0 \\ 0 & \rho \end{bmatrix};\quad 10^{-4} \le \rho \le 10^{-6};$$
$$0.97 \le \alpha(k) \le 1;\quad 10^3 \le P(0) \le 10^6.\tag{1.26}$$

The matrix $\Omega(\cdot)$ had a dimension $(L+1) \times N_w$, whereas the second row had only one unity element (the others were zero). The position of that element was computed by:

$$i = k \bmod (N_w) + 1;\quad k > N_w\tag{1.27}$$

Detailed derivation of the recursive L-M learning algorithm is given in the Appendix 1. After this, the given up topology and learning will be incorporated in the T-S identification and control rules and applied for an anaerobic wastewater distributed parameter decentralized system identification and control in each collocation point.

1.3 Description of the Decentralized Direct I-Term Fuzzy-Neural Multi-Model Control System

The block-diagrams of the complete control system and its identification and control parts are schematically depicted in Figs. 1.3, 1.4 and 1.5. The structure of the entire control system [26, 27] contained Fuzzifier, Fuzzy Rule-Based Inference System (FRBIS), and defuzzifier.

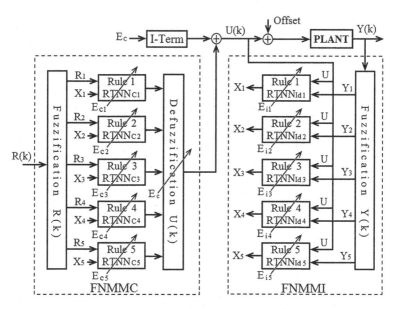

Fig. 1.3 Block diagram of the direct decentralized HFNMM control system with I-term

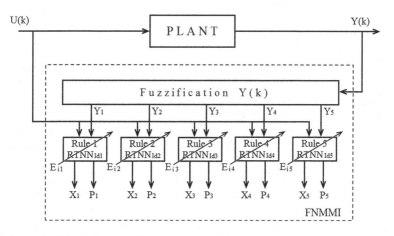

Fig. 1.4 Detailed block-diagram of the FNMM identifier

The FRBIS contained five identification and five feedback-feedforward (FB-FF) control T-S fuzzy rules (see Figs. 1.3, 1.4, 1.5 for more details). The plant output variables and its correspondent reference variables depended on space and time. They are fuzzified on space and represented by five membership functions which centers are the five collocation points of the plant (four points for the fixed bed and one point for the recirculation tank). The main objective of the Fuzzy-Neural Multi-Model Identifier (FNMMI), containing five rules, is to issue states

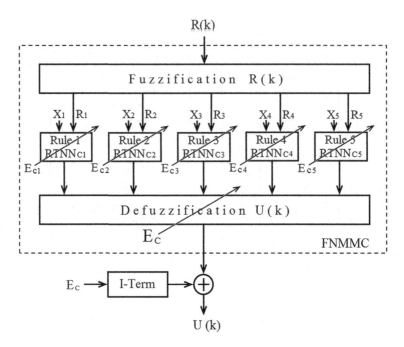

Fig. 1.5 Detailed block-diagram of the direct decentralized HFNMM controller with I-term

and parameters for the direct adaptive Fuzzy-Neural Multi-Model Feedback Controller (FNMMFBC) when the FNMMI outputs follows the outputs of the plant in the five measurement (collocation) points with minimum error of approximation. The control part of the system is a direct adaptive Fuzzy-Neural Multi-Model Controller (FNMMC). The objective of the direct adaptive FNMM controller, containing five Feedback-Feedforward (FB-FF) T-S control rules, is to speed up the reaction of the control system, and to augment the resistance of the control system to process and measurement noises, reducing the error of control, so that the plant outputs in the five measurement points tracked the corresponding reference variables with minimum error of tracking. The upper hierarchical level of the FNMM control system is one- layer- perceptron which represented the defuzzifier, [26, 27]. Its output is summed with an I-term control to form the total control. The hierarchical FNMM controller has two levels—Lower Level of Control (LLC), and Upper Level of Control (ULC). It is composed of three parts (see Fig. 1.4): (1) Fuzzification, where the normalized reference vector signal contained reference components of five measurement points; (2) Lower Level Inference Engine, which contained ten T-S fuzzy rules (five rules for identification and five rules for FB-FF control), operating in the corresponding measurement points; (3) Upper Hierarchical Level of neural defuzzification. The detailed block-diagram of the FNMMI (see Fig. 1.3), contained a space plant output fuzzifier and five identification T-S fuzzy rules, labeled as RI_i, which consequent parts are

RTNN learning procedures, [26, 27]. The identification T-S fuzzy rules have the form:

RI_i: *If* $x(k)$ *is* A_i *and* $u(k)$ *is* B_i *then* $Y_i = \Pi_i(L, M, N_i, Y_{di}, U, X_i, A_i, B_i, C_i, E_i)$, $i = 1 - 5$

$$(1.28)$$

The detailed block-diagram of the FNMMC, given on Fig. 1.4, contained a spaced plant reference fuzzifier and five control T-S fuzzy rules (five FB-FF control rules), which consequent FB-FF parts are also RTNN learning procedures, [27], using the state information, issued by the corresponding identification rules. The consequent part of each FB-FF control rule (the consequent learning procedure) has the M, L, N_i RTNN model dimensions, R_i, Y_{di}, X_i, E_{ci} inputs and U_{ffi}, outputs used by the defuzzifier. The T-S fuzzy rule has the form:

RCi: *If* $R(k)$ *is* B_i *then* $Uc_i = \Pi_i(M, L, N_i, R_i, Y_{di}, X_i, X_{ci}, J_i, B_i, C_i, E_{ci})$, $i = 1 - 5$

$$(1.29)$$

The I-term control algorithm is as follows:

$$UI(k + 1) = UI(k) + To\,K(k)E_c(k), \qquad (1.30)$$

where: To is period of discretization and K is the I-term gain. An appropriate choice for the I-term gain K is a proportion of the inverse input/output plant gain, i.e.:

$$K(k) = \eta\,(C\,B)^+. \qquad (1.31)$$

The product of the pseudo-inverse $(C\,B)^+$ by the output error $E_c(k)$ transformed the output error in input error which equates the dimensions in the equation of the I-term control. The T-S rule, generating the I-term part of the control executed both Eqs. (1.30) and (1.31), representing a computational procedure, given by:

RCI : *If* Y_d *is A then* $UI = \Pi(M, L, B, C, E_c, To, \eta)$, $\qquad (1.32)$

The total control is a sum of the defuzzifier output, and the I-term control, as:

$$U(k) = U_c(k) + U_I(k), \qquad (1.33)$$

The total control (1.33) is generated by the procedure incorporated in the T-S rule:

$$RC: \textbf{\textit{If}} \ Y \ \textit{is} \ A \ \textbf{\textit{then}} \ U - \Pi(M, U_c, U_I), \qquad (1.34)$$

The defuzzification learning procedure, which correspond to the single layer perceptron learning is described by:

$$Uc = \Pi(M, L, N, R, Y_d, Uo, X, A, B, C, E) \qquad (1.35)$$

The T-S rule and the defuzzification of the plant output of the fixed bed with respect to the space variable z (λi, z is the correspondent membership function), are given by:

$$RO_i: \textbf{\textit{If}} \ Y_{i,t} \ \textit{is} \ A_i \ \textbf{\textit{then}} \ Y_{i,t} = a_i^T Y_t + b_i, \ i = 1, 2, 3, 4; \qquad (1.36)$$

$$Y_z = \left[\sum_i \gamma_{i,z} a_i^T \right] Y_t + \Sigma_i \gamma_{i,z} b_i \ ; \gamma_{i,z} = \lambda_{i,z} / \left(\Sigma_j \lambda_{j,z} \right). \qquad (1.37)$$

The direct adaptive neural control algorithm, which appeared in the consequent part of the local fuzzy control rule RCi (1.29) is a FB-FF control, using the states issued by the correspondent identification local fuzzy rule RIi (1.28).

1.4 Description of the Decentralized Indirect (Sliding Mode) I-Term Fuzzy-Neural Multi-Model Control System

The block-diagram of the indirect FNMM control system is given on Fig. 1.6. It contained a FNMM identifier and a Sliding Mode (SM) FNMM controller, depicted in Figs. 1.4 and 1.7.

The structure of the entire control system, [26], contained Fuzzifier, Fuzzy Rule-Based Inference System, containing ten T-S fuzzy rules (five identification, five sliding mode control, defuzzifier, one I-term control and one total control rules). Due to the learning abilities of the defuzzifier, the exact form of the control membership functions is not need to be known. The plant output variable and its correspondent reference variable depended on space and time, and they are fuzzified on space. The membership functions of the fixed-bed output variables are triangular or trapezoidal ones and that—belonging to the output variables of the recirculation tank are singletons. Centers of the membership functions are the respective collocation points of the plant.

The main objective of the FNMM Identifier (FNMMI) (see Fig. 1.4), containing five T-S rules, is to issue states and parameters for the indirect adaptive FNMM Controller (FNMMC) when the FNMMI outputs follows the outputs of the plant in the five measurement (collocation) points with minimum MSE of approximation.

The objective of the indirect adaptive FNMM controller, containing five Sliding Mode Control (SMC) rules, defuzzifier, one I-term, and one total control rules is to

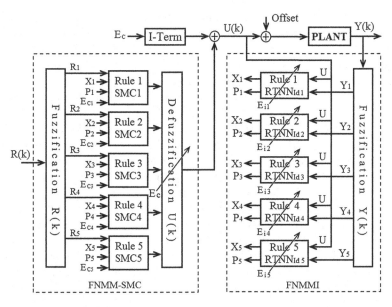

Fig. 1.6 Block diagram of the indirect (SM) decentralized HFNMM control system with I-term

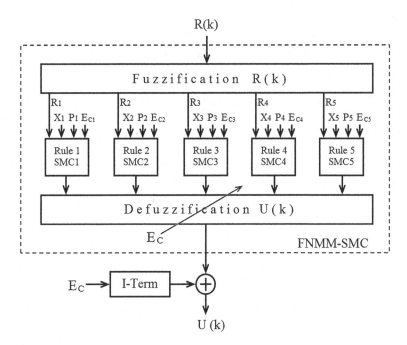

Fig. 1.7 Detailed block-diagram of the indirect (SM) decentralized HFNMM controller with I-term

reduce the error of control, so that the plant outputs of the four measurement points tracked the corresponding reference variables with minimum MSE. The hierarchical FNMM controller (see Fig. 1.7) has two levels—Lower Level of Control (LLC), and Upper Level of Control (ULC). It is composed of three parts: (1) Fuzzification, where the normalized reference vector signal contained reference components of five measurement points; (2) Lower Level Inference Engine, which contained ten T-S fuzzy rules (five rules for identification, five rules for SM control), plus one rule for I-term control, and one rule for total control. The five identification and five SM control rules operate in the corresponding measurement points; (3) Upper Hierarchical Level of neural defuzzification, represented by one layer perceptron, [26]. The block-diagram of the FNMMI (see Fig. 1.4), contained a space plant output fuzzifier and five identification T-S fuzzy rules RI_i, which consequent parts are learning procedures, [26], given by (1.28).

The block-diagram of the FNMMC (see Fig. 1.7) contained a spaced plant reference fuzzifier, five SMC T-S fuzzy rules, defuzzifier, one T-term control, and one total control T-S fuzzy rules. The consequent parts of the SMC T-S fuzzy rules are SMC procedures, [26]. Using the state, and parameter information, issued by the corresponding identification T-S rules, we could write the SMC T-S fuzzy rules as:

$$RC_i : \textit{If } R(k) \textit{ is } C_i \textit{ then } U_i = \Pi_i(M, L, N_i, R_i, Y_{di}, X_i, A_i, B_i, C_i, E_{ci}), i = 1 - 5$$
$$(1.38)$$

The I-term control algorithm, given by (1.30) (1.31), is incorporated in the T-S I-term control rule (1.32) as a computational procedure of the antecedent part. The total control is a sum of the defuzzifier output, and the I-term, formulated in the same manner as in the previous paragraph, i.e. (1.33) (1.34).

The defuzzification learning procedure, which correspond to the single layer perceptron learning is described also in the same manner as (1.35). The T-S rule and the defuzzification of the plant output of the fixed bed with respect to the space variable z ($\lambda_{i,z}$ is the correspondent membership function), are given by (1.36) (1.37). The indirect (SMC) adaptive neural control algorithm, which appeared in the consequent part of the local fuzzy control rule RCi (1.38) is a Feedback-Feedforward (FB-FF) SM control, using the parameters and states issued by the correspondent identification local fuzzy rule RIi (1.28).

1.4.1 Sliding Mode Control Systems Design

Here the Indirect Adaptive Neural Control (IANC) algorithm, which appeared in the consequent part of the local fuzzy control rule RCi (1.38) is viewed as a Sliding Mode Control (SMC), [26], designed using the parameters and states issued by the correspondent identification local fuzzy rule RIi (1.28), approximating the plant in the corresponding collocation point.

Let us suppose that the studied local nonlinear plant model possess the following structure:

$$X_p(k+1) = F[X_p(k), U(k)]; \ Y_p(k) = G[X_p(k)] \tag{1.39}$$

where: $X_p(k)$, $Y_p(k)$, $U(k)$ are plant state, output and input vector variables with dimensions Np, L and M, where L > M (rectangular system) is supposed; F and G are smooth, odd, bounded nonlinear functions. The linearization of the activation functions of the local learned identification RTNN model, which approximates the plant leads to the following linear local plant model:

$$X(k+1) = AX(k) + BU(k); \quad Y(k) = CX(k) \tag{1.40}$$

where L > M (rectangular system), is supposed. Let us define the following sliding surface with respect to the output tracking error:

$$S(k+1) = E(k+1) + \sum_{i=1}^{P} \gamma_i E(k-i+1); \quad |\gamma_i| < 1; \tag{1.41}$$

where: $S(\cdot)$ is the sliding surface error function; $E(\cdot)$ is the systems local output tracking error; γ_i are parameters of the local desired error function; P is the order of the error function. The additional inequality in (1.41) is a stability condition, required for the sliding surface error function. The local tracking error is defined as $E(k) = R(k) - Y(k)$, where: $R(k)$ is a L-dimensional local reference vector and $Y(k)$ is an local output vector with the same dimension. The objective of the sliding mode control systems design is to find a control action which maintains the systems error on the sliding surface assuring that the output tracking error reached zero in P steps, where P < N, which is fulfilled if $S(k + 1) = 0$. As the local approximation plant model (1.40), is controllable, observable and stable, [30, 32, 33], the matrix A is block-diagonal, and L > M (rectangular system is supposed), the matrix product (CB) is nonsingular with rank M, and the plant states $X(k)$ are smooth non-increasing functions. Now, from (1.40, 1.41), taking into account the mentioned above observations, it is easy to obtain the equivalent control capable to lead the system to the sliding surface which yields:

$$U_{eq}(k) = (CB)^+ \left[-CAX(k) + R(k+1) + \sum_{i=1}^{P} \gamma_i E(k-i+1) \right] + Of \tag{1.42}$$

$$(CB)^+ = \left[(CB)^T (CB) \right]^{-1} (CB)^T. \tag{1.43}$$

Here the added offset Of is a learnable M-dimensional constant vector which is learnt using a simple delta rule (see, [1] for more details), where the error of the plant input is obtained backpropagating the output error through the adjoint RTNN model. An easy way for learning the offset is using the following delta rule where the input error is obtained from the output error multiplying it by the same pseudoinverse matrix, as it is:

$$Of(k + 1) = Of(k + 1) = Of(k) + \eta(CB)^+ E(k). \qquad (1.44)$$

If we compare the I-term expression (1.30, 1.31) with the Offset learning (1.44) we could see that they are equal which signified that the I-term generate a compensation offset capable to eliminate steady state errors caused by constant perturbations and discrepancies in the reference tracking caused by non equal input/output variable dimensions (rectangular case systems). So introducing an I-term control it is not necessary to use a compensation offset in the SM control law (1.42). The SMC avoiding chattering is taken using a saturation function inside a bounded control level Uo, taking into account plant uncertainties. The proposed SMC cope with the characteristics of the wide class of plant model reduction neural control with reference model, and represents an indirect adaptive neural control, given by Baruch, [26, 30].

1.5 Analytical Model of the Anaerobic Digestion Bioprocess Plant

The anaerobic digestion systems block diagram is depicted on Fig. 1.8. It consists of fixed bed reactor and a recirculation tank. The physical meaning of all variables and constants (also its values), are summarized on Table 1.1.

The complete analytical model of wastewater treatment anaerobic bioprocess, taken from [26–28], could be described by the following system of PDE and ODE (also for the recirculation tank):

$$\frac{\partial X_1}{\partial t} = (\mu_1 - \varepsilon D)X_1, \quad \mu_1 = \mu_{1\max} \frac{S_1}{K'_{s_1} X_1 + S_1}, \qquad (1.45)$$

$$\frac{\partial X_2}{\partial t} = (\mu_2 - \varepsilon D)X_2, \quad \mu_2 = \mu_{2s} \frac{S_2}{K'_{s_2} X_2 + S_2 + \frac{S_2^2}{K_{I_2}}}, \qquad (1.46)$$

$$\frac{\partial S_1}{\partial t} = \frac{E_z}{H^2} \frac{\partial^2 S_1}{\partial z^2} - D\frac{\partial S_1}{\partial t} - k_1 \mu_1 X_1, \qquad (1.47)$$

$$\frac{\partial S_2}{\partial t} = \frac{E_z}{H^2} \frac{\partial^2 S_2}{\partial z^2} - D\frac{\partial S_2}{\partial t} + k_2 \mu_1 X_1 - k_3 \mu_2 X_2, \qquad (1.48)$$

$$S_1(0, t) = \frac{S_{1,in}(t) + RS_{1T}}{R + 1}, \quad S_2(0, t) = \frac{S_{2,in}(t) + RS_{2T}}{R + 1}, \quad R = \frac{Q_T}{DV_{eff}}, \qquad (1.49)$$

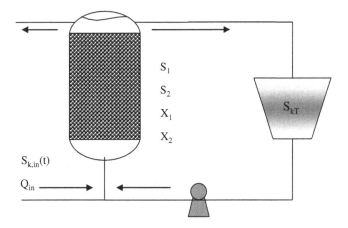

Fig. 1.8 Block-diagram of anaerobic digestion bioreactor

$$\frac{\partial S_1}{\partial z}(1,t) = 0, \quad \frac{\partial S_2}{\partial z}(1,t) = 0. \tag{1.50}$$

For practical purpose, the full PDE anaerobic digestion process model, [28], could be reduced to an ODE system using an early lumping technique and the Orthogonal Collocation Method (OCM), [29], in four points (0.2H, 0.4H, 0.6H, 0.8H) obtaining the following system of OD equations:

$$\frac{dX_{1,i}}{dt} = \left(\mu_{1,i} - \varepsilon D\right)X_{1,i}, \quad \frac{dX_{2,i}}{dt} = \left(\mu_{2,i} - \varepsilon D\right)X_{2,i}, \tag{1.51}$$

$$\frac{dS_{1,i}}{dx} = \frac{E_z}{H^2}\sum_{j=1}^{N+2} B_{i,j}S_{1j} - D\sum_{j=1}^{N+2} A_{i,j}S_{1j} - k_1\mu_{1,i}X_{1,i}, \tag{1.52}$$

$$\frac{dS_{1T}}{dt} = \frac{Q_T}{V_T}\left(S_1(1,t) - S_{1T}\right), \quad \frac{dS_{2T}}{dt} = \frac{Q_T}{V_T}\left(S_2(1,t) - S_{2T}\right). \tag{1.53}$$

$$\frac{dS_{2,i}}{dx} = \frac{E_z}{H^2}\sum_{j=1}^{N+2} B_{i,j}S_{2j} - D\sum_{j=1}^{N+2} A_{i,j}S_{2j} + k_2\mu_{1,i}X_{1,i} - k_3\mu_{2,i}X_{2,i}, \tag{1.54}$$

$$\frac{dS_{1T}}{dt} = \frac{Q_T}{V_T}\left(S_{1,N+2} - S_{1T}\right), \quad \frac{dS_{2T}}{dt} = \frac{Q_T}{V_T}\left(S_{2,N+2} - S_{2T}\right), \tag{1.25}$$

$$S_{k,1} = \frac{1}{R+1}S_{k,in}(t) + \frac{R}{R+1}S_{kT}, \quad S_{k,N+2} = \frac{K_1}{R+1}S_{k,in}(t) + \frac{K_1 R}{R+1}S_{kT} + \sum_{i=2}^{N+1} K_i S_{k,i} \tag{1.56}$$

Table 1.1 Summary of the variables in the plant model

Variable	Units	Name	Value
z	$z{\in}[0,1]$	Space variable	
t	D	Time variable	
E_z	m^2/d	Axial dispersion coefficient	1
D	1/d	Dilution rate	0.55
H	m	Fixed bed length	3.5
X_1	g/L	Concentration of acidogenic bacteria	
X_2	g/L	Concentration of methanogenic bacteria	
S_1	g/L	Chemical oxygen demand	
S_2	mmol/L	Volatile fatty acids	
ε		Bacteria fraction in the liquid phase	0.5
k_1	g/g	Yield coefficients	42.14
k_2	mmol/g	Yield coefficients	250
k_3	mmol/g	Yield coefficients	134
μ_1	1/d	Acidogenesis growth rate	
μ_2	1/d	Methanogenesis growth rate	
μ_{1max}	1/d	Maximum acidogenesis growth rate	1.2
μ_{2s}	1/d	Maximum methanogenesis growth rate	0.74
K_{1s}	g/g	Kinetic parameter	50.5
K_{2s}	mmol/g	Kinetic parameter	16.6
K_{I2}	mmol/g	Kinetic parameter	256
Q_T	m^3/d	Recycle flow rate	0.24
V_T	m^3	Volume of the recirculation tank	0.2
S_{1T}	g/L	Concentration of chemical oxygen demand in the recirculation tank	
S_{2T}	mmol/L	Concentration of volatile fatty acids in the recirculation tank	
Q_{in}	m^3/d	Inlet flow rate	0.31
V_B	m^3	Volume of the fixed bed	1
V_{eff}	m^3	Effective volume tank	0.95
$S_{1,\,in}$	g/l	Inlet substr. concentration	
$S_{2,\,in}$	mmol/L	Inlet substr. concentration	

$$K_1 = -\frac{A_{N+2,1}}{A_{N+2,N+2}}, \quad K_i = -\frac{A_{N+2,i}}{A_{N+2,N+2}}, \tag{1.57}$$

$$A = \Lambda\phi^{-1}, \quad \Lambda = [\varpi_{m,l}], \quad \varpi_{m,l} = (l-1)z_m^{l-2}, \tag{1.58}$$

$$B = \Gamma\phi^{-1}, \quad \Gamma = [\tau_{m,l}], \quad \tau_{m,l} = (l-1)(l-2)z_m^{l-3}, \quad \phi_{m,l} = z_m^{l-1} \tag{1.59}$$

$$i = 2,\ldots,N+1, \quad m,l = 1,\ldots,N+2. \tag{1.60}$$

The reduced plant model (1.51)–(1.60) (here (1.52) represented the OD equations of the recirculation tank), [28], could be used as unknown plant model which generate input/output process data for decentralized adaptive Fuzzy Neural Multi Model (FNMM) control system design and control systems simulation.

1.6 Simulation Results

In this paragraph, graphical and numerical simulation results of fuzzy-neural system identification, direct and indirect decentralized adaptive fuzzy-neural control of DPS bioprocess plant, will be given. For lack of space we will give complete graphical results only for the most representative X_1, X_2 variables of the fixed bed and the variables S_{1T}, S_{2T} of the recirculation tank. This part shows graphical and numerical simulation results obtained using the model of the wastewater treatment anaerobic digestion bioprocess plant as an input/output data generator.

1.6.1 Simulation Results of the System Identification

The decentralized FNMM identifier used a set of five T-S fuzzy. The RTNN topology is given by the Eqs. (1.1)–(1.5), the BP RTNN learning is given by (1.6)–(1.13), and the L-M RTNN learning is given by (1.14)–(1.27). The topology of the first four RTNNs is (2-6-4) (2 inputs, 6 neurons in the hidden layer, 4 outputs) and the last one has topology (2-6-2), corresponding to the fixed bed plant behavior in each collocation point and the recirculation tank. The RTNNs identified the following fixed bed variables: X_1 (acidogenic bacteria), X_2 (methanogenic bacteria), S_1 (chemical oxygen demand) and S_2 (volatile fatty acids), in the following four collocation points, $z = 0.2H$, $z = 0.4H$, $z = 0.6H$, $z = 0.8H$, and the following variables in the recirculation tank: S_{1T} (chemical oxygen demand) and S_{2T} (volatile fatty acids). The graphical simulation results of RTNNs BP and L-M learning are obtained on-line during 500 iterations. The learning rate parameters of RTNN have small values which are different for the different measurement point variables. Figures 1.9, 1.10, 1.11, 1.12, 1.13, 1.14 and 1.15 showed graphical simulation results of open loop decentralized plant identification. Figures 1.9, 1.11 and 1.13a, b gives the T-S fuzzy-neural approximation of the correspondent variables for 500 iterations of the L-M learning. Figures 1.10, 1.12 and 1.13c, d compared the time-dependent graphics of the correspondent plant output variables with the correspondent RTNNs outputs only at for the first 10 iterations of L-M learning. Figures 1.14 and 1.15 show the 3-D plot of X_1, X_2 during L-M learning. The input signals applied are:

$$S_{1,in} = 6 + 2 \sin\left(\frac{4\pi t}{100}\right) + \sin\left(\frac{2\pi t}{100}\right) + 3\cos\left(\frac{\pi t}{100}\right) \tag{1.61}$$

$$S_{2,in} = 50 + 30 \sin\left(\frac{2\pi t}{100}\right) + 15 \sin\left(\frac{6\pi t}{100}\right) + 15\cos\left(\frac{3\pi t}{100}\right) \tag{1.62}$$

The MSE of the decentralized FNMM approximation of plant variables in the collocation points, using the L-M and BP RTNN learning are shown in Tables 1.2 and 1.3. The graphical y numerical results of decentralized FNMM identification (see Figs. 1.9, 1.10, 1.11, 1.12, 1.13, 1.14, 1.15; Tables 1.2, 1.3) showed a good

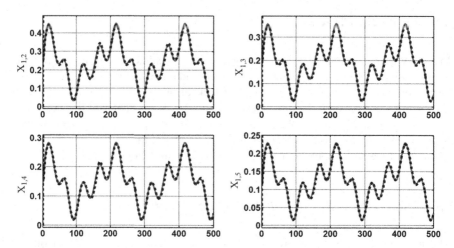

Fig. 1.9 Graphical simulation results of the FNMM identification of X_1 in z = 0.2H; 0.4H; 0.6H; 0.8H (acidogenic bacteria in the corresponding fixed bed points—$X_{1,2}$; $X_{1,3}$; $X_{1,4}$; $X_{1,5}$) by four fuzzy rules (*dotted line*—RTNN output, *continuous line*—plant output) for 500 iteration of L-M RTNN learning

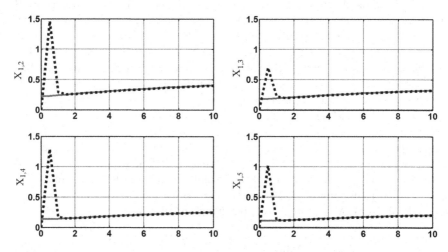

Fig. 1.10 Detailed graphical simulation results of the FNMM identification of X_1 in z = 0.2H; 0.4H; 0.6H; 0.8H (acidogenic bacteria in the corresponding fixed bed points—$X_{1,2}$; $X_{1,3}$; $X_{1,4}$; $X_{1,5}$) by four fuzzy rules RTNNs (*dotted line*—RTNN output, *continuous line*—plant output) for the first 10 iterations of the L-M RTNN learning

HFNMMI convergence and precise plant output tracking (MSE 1.0842 for the L-M, and 1.2423 for the BP RTNN learning in the worse case). Next, some results of direct and indirect decentralized hierarchical fuzzy-neural multi-model control with I-term and L-M learning will be given.

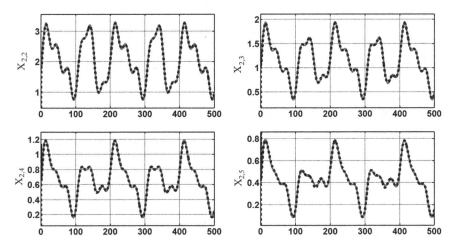

Fig. 1.11 Graphical simulation results of the FNMM identification of X_2 in z = 0.2H; 0.4H; 0.6H; 0.8H (methanogenic bacteria in the corresponding fixed bed points—$X_{2,2}$; $X_{2,3}$; $X_{2,4}$; $X_{2,5}$) by four fuzzy rules RTNNs (*dotted line*—RTNN output, *continuous line*—plant output) for 500 iteration of L-M RTNN learning

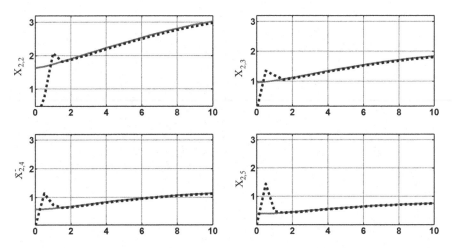

Fig. 1.12 Detailed graphical simulation results of the FNMM identification of X_2 in z = 0.2H; 0.4H; 0.6H; 0.8H (methanogenic bacteria in the corresponding fixed bed point—$X_{2,2}$; $X_{2,3}$; $X_{2,4}$; $X_{2,5}$) by four fuzzy rules RTNNs (*dotted line*—RTNN output, *continuous line*—plant output) for the first 10 iterations of the L-M RTNN learning

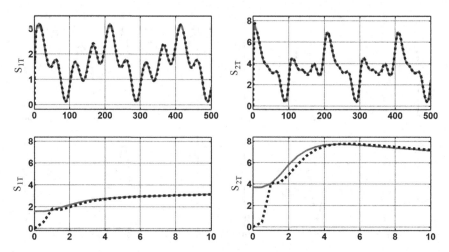

Fig. 1.13 Graphical simulation results and detailed graphical simulation results of the FNMM identification of S_{1T} (chemical oxygen demand in the recirculation tank) and S_{2T} (volatile fatty acids in the recirculation tank) (*dotted line*—RTNN output, *continuous line*—plant output) for 500 (S_{1T}; S_{2T}) and for 10 (S_{1T}; S_{2T}) iteration of L-M RTNN learning

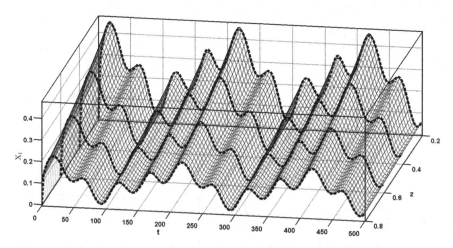

Fig. 1.14 Graphics of the 3D view of X_1 space/time approximation during its L-M RTNN learning in four points

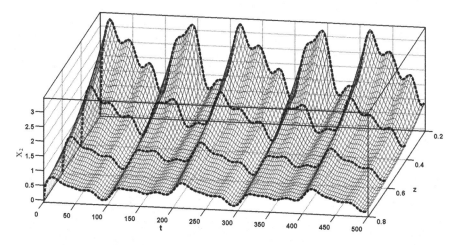

Fig. 1.15 Graphics of the 3D view of X_2 space/time approximation during its L-M RTNN learning in four points

Table 1.2 MSE of the decentralized FNMM approximation of the bioprocess output variables in the collocation points, using the L-M RTNN learning

Collocation point	X_1	X_2	S_1/S_{1T}	S_2/S_{2T}
z = 0.2	0.0016	0.0043	0.0386	1.0842
z = 0.4	0.0003	0.0013	0.0196	0.3151
z = 0.6	0.0014	0.0007	0.0091	0.1103
z = 0.8	0.0008	0.0013	0.0069	0.0412
Recirculation tank			0.0053	0.0295

Table 1.3 MSE of the decentralized FNMM approximation of the bioprocess output variables in the collocation points, using the BP RTNN learning

Collocation point	X_1	X_2	S_1/S_{1T}	S_2/S_{2T}
z = 0.2	0.0041	0.0098	0.1401	1.2423
z = 0.4	0.0011	0.0037	0.1459	0.6394
z = 0.6	0.0004	0.0027	0.1358	0.3920
z = 0.8	0.0017	0.0021	0.0683	0.1941
Recirculation tank			0.0377	0.1094

1.6.2 Simulation Results of the Direct HFNMM Control with and without I-Term

Figures 1.16, 1.17, 1.18, 1.19, 1.20, 1.21, 1.22, 1.23, 1.24 and 1.25 show the graphical simulation results of the direct decentralized HFNMM control with and without I-term, where the outputs of the plant are compared with the reference

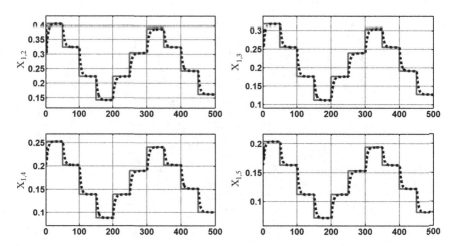

Fig. 1.16 Graphical simulation results of the direct decentralized HFNMM I-term control of X_1 (acidogenic bacteria in the fixed bed) (*dotted line*—plant output, *continuous*-reference) in four collocation points (0.2H, 0.4H, 0.6H, 0.8H) for 500 iterations of L-M RTNN learning ($X_{1,2}$; $X_{1,3}$; $X_{1,4}$; $X_{1,5}$)

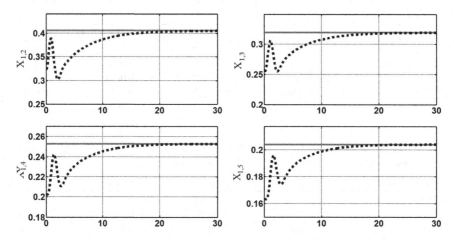

Fig. 1.17 Detailed graphical simulation results of the direct decentralized HFNMM I-term control of X_1 (acidogenic bacteria in the fixed bed) (*dotted line*—plant output, *continuous*-reference) in four collocation points (0.2H, 0.4H, 0.6H, 0.8H) for the first 30 iterations L-M RTNN learning ($X_{1,2}$; $X_{1,3}$; $X_{1,4}$; $X_{1,5}$)

signals. The reference signals are train of pulses with uniform duration and random amplitude. The MSE of control for each output signal and each measurement point are given on Tables 1.4, 1.5.

For sake of comparison the MSE of direct decentralized HFNMM proportional control (without I-term) for each output signal and each measurement point are given on Table 1.5.

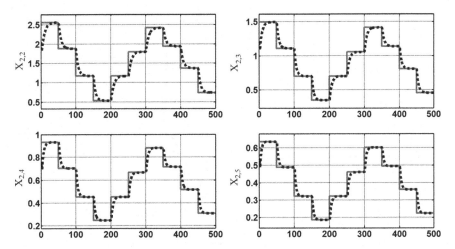

Fig. 1.18 Graphical simulation results of the direct decentralized HFNMM I-term control of X_2 (methanogenic bacteria in the fixed bed) (*dotted line*—plant output, *continuous*-reference) in four collocation points (0.2H, 0.4H, 0.6H, 0.8H) for 500 iterations of L-M RTNN learning ($X_{2,2}$; $X_{2,3}$; $X_{2,4}$; $X_{2,5}$)

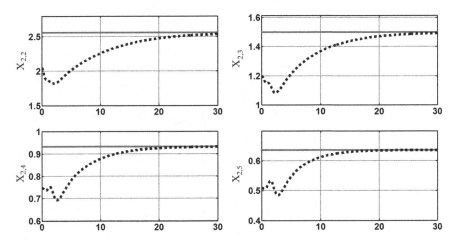

Fig. 1.19 Detailed graphical results of the direct decentralized HFNMM I-term control of X_2 (methanogenic bacteria in the fixed bed) (*dotted line*—plant output, *continuous*-reference) in four collocation points (0.2H, 0.4H, 0.6H, 0.8H) for the first 30 iterations of L-M learning ($X_{2,2}$; $X_{2,3}$; $X_{2,4}$; $X_{2,5}$)

For sake of comparison, graphical results of direct decentralized HFNMM proportional control (without I-term) for the X_1 variable are presented on Figs. 1.23, 1.24, 1.25. The results show that the proportional control could not eliminate the static error due to inexact approximation and constant process or measurement disturbances.

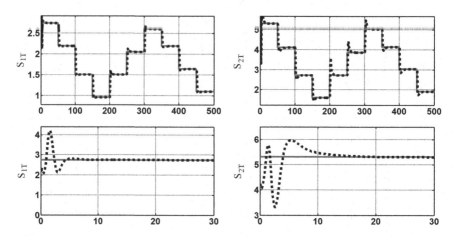

Fig. 1.20 Graphical simulation results and detailed graphical simulation results of the direct HFNMM I-term control of S_{1T} (chemical oxygen demand in the recirculation tank) and S_{2T} (volatile fatty acids in the recirculation tank) (*dotted line*—RTNN output, *continuous line*—plant output) for 500 (S_{1T}; S_{2T}) and for 30 (S_{1T}; S_{2T}) iterations of L-M RTNN learning

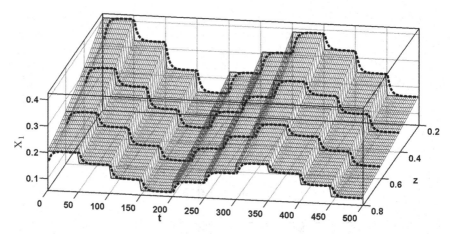

Fig. 1.21 Graphics of the 3D view of X_1 space/time approximation and direct decentralized HFNMM I-term control in four collocation points of the fixed bed

The graphical y numerical results of direct decentralized HFNMM I-term control showed a good reference tracking (MSE is of 0.7166 for the I-term control and 16.69 for the control without I-term in the worse case). The results showed that the I-term control eliminated constant disturbances and approximation errors, and the proportional control could not.

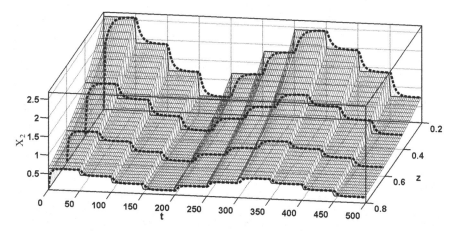

Fig. 1.22 Graphics of the 3D view of X_2 space/time approximation and direct decentralized HFNMM I-term control in four collocation points of the fixed bed

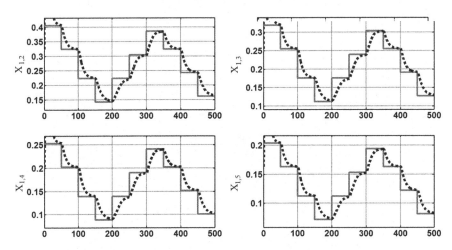

Fig. 1.23 Graphical simulation results of the direct decentralized HFNMM proportional control (without I-term) of X_1 (acidogenic bacteria in the fixed bed) (*dotted line*—plant output, *continuous*-reference) in four collocation points (0.2H, 0.4H, 0.6H, 0.8H) for 500 iterations of L-M RTNN learning ($X_{1,2}$; $X_{1,3}$; $X_{1,4}$; $X_{1,5}$)

1.6.3 Simulation Results of the Indirect HFNMM I-Term SMC

Figures 1.26, 1.27, 1.28, 1.29, 1.30, 1.31, 1.32, 1.33, 1.34 and 1.35 show graphical simulation results of the indirect (sliding mode) decentralized HFNMM control with and without I-term. The MSE of control for each output signal and each measurement point are given on Tables 1.6, 1.7.

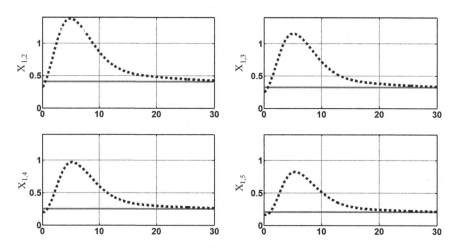

Fig. 1.24 Detailed graphical results of the direct decentralized HFNMM proportional control (without I-term) of X_1 (acidogenic bacteria in the fixed bed) (*dotted line*—plant output, *continuous*-reference) in four collocation points (0.2H, 0.4H, 0.6H, 0.8H) for the first 30 iterations of learning ($X_{1,2}$; $X_{1,3}$; $X_{1,4}$; $X_{1,5}$)

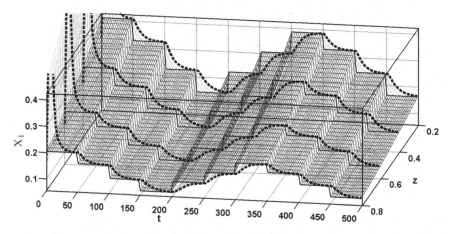

Fig. 1.25 Graphics of the 3D view of X_1 space/time approximation and direct decentralized HFNMM proportional control (without I-term) in four collocation points of the fixed bed

Table 1.4 MSE of the direct decentralized HFNMM I-term control of the bioprocess plant

Collocation point	X_1	X_2	S_1/S_{1T}	S_2/S_{2T}
$z = 0.2$	0.0002	0.0185	0.1055	0.7166
$z = 0.4$	0.0001	0.0051	0.0470	0.1965
$z = 0.6$	0.0001	0.0016	0.0238	0.0784
$z = 0.8$	0.0001	0.0007	0.0141	0.0426
Recirculation tank			0.0067	0.0275

Table 1.5 MSE of the direct decentralized HFNMM proportional control (without I-term) of the bioprocess plant

Collocation point	X_1	X_2	S_1/S_{1T}	S_2/S_{2T}
$z = 0.2$	0.0062	0.0247	7.2037	7.4883
$z = 0.4$	0.0044	0.0128	5.3024	11.6640
$z = 0.6$	0.0031	0.0080	3.9731	15.5562
$z = 0.8$	0.0023	0.0055	3.1020	2.3535
Recirculation tank			4.7069	16.6883

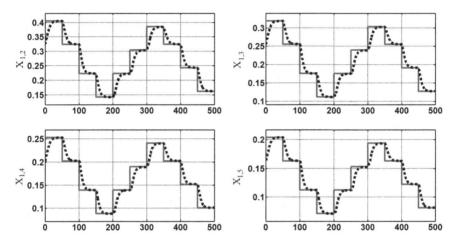

Fig. 1.26 Graphical simulation results of the indirect (SMC) decentralized HFNMM I-term control of X_1 (acidogenic bacteria in the fixed bed) (*dotted line*—plant output, *continuous*—reference) in four collocation points (0.2H, 0.4H, 0.6H, 0.8H) for 500 iterations of L-M ($X_{1,2}$; $X_{1,3}$; $X_{1,4}$; $X_{1,5}$)

The reference signals are train of pulses with uniform duration and random amplitude and the outputs of the plant are compared with the reference signals. The graphical y numerical results of the indirect (sliding mode) decentralized control showed precise reference tracking (MSE is 0.2755 in the worse case).

The comparison of the direct and indirect decentralized control with I-term showed a good results for both control methods (see Tables 1.4, 1.6) with slight priority for the indirect control (0.7166 vs. 0.2755) due to its better plant dynamics compensation ability and adaptation.

For sake of comparison, graphical results of indirect decentralized HFNMM proportional control (without I-term) for the X_1 variable are presented on Figs. 1.33, 1.34, 1.35. The results show that the proportional control could not eliminate the static error caused of inexact approximation and constant process or measurement disturbances.

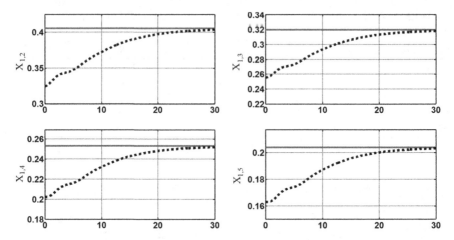

Fig. 1.27 Detailed graphical results of the indirect (SMC) decentralized HFNMM I-term control of X_1 (acidogenic bacteria in the fixed bed) (*dotted line*—plant output, *continuous*-reference) in four collocation points (0.2H, 0.4H, 0.6H, 0.8H) for the first 30 iterations of L-M ($X_{1,2}$; $X_{1,3}$; $X_{1,4}$; $X_{1,5}$)

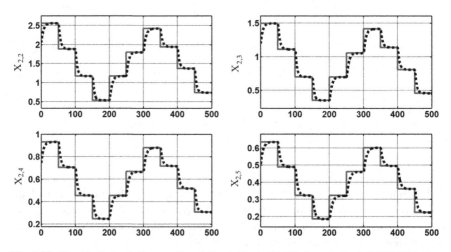

Fig. 1.28 Graphical simulation results of the indirect (SMC) decentralized HFNMM I-term control of X_2 (methanogenic bacteria in the fixed bed) (*dotted line*—plant output, *continuous*-reference) in four collocation points (0.2H, 0.4H, 0.6H, 0.8H) for 500 iterations of L-M ($X_{2,2}$; $X_{2,3}$; $X_{2,4}$; $X_{2,5}$)

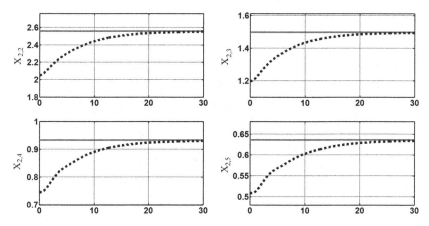

Fig. 1.29 Detailed graphical results of the indirect (SMC) decentralized HFNMM I-term control of X_2 (methanogenic bacteria in the fixed bed) (*dotted line*—plant output, *continuous*-reference) in four collocation points (0.2H, 0.4H, 0.6H, 0.8H) for the first 30 iterations of L-M ($X_{2,2}$; $X_{2,3}$; $X_{2,4}$; $X_{2,5}$)

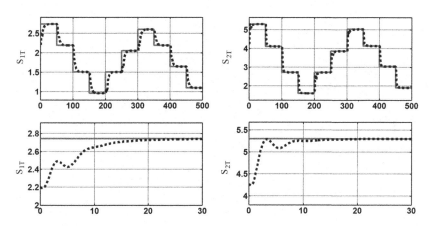

Fig. 1.30 Graphical simulation results and detailed graphical simulation results of the indirect (SMC) HFNMM I-term control of S_{1T} (chemical oxygen demand in the recirculation tank) and S_{2T} (volatile fatty acids in the recirculation tank) (*dotted line*—RTNN output, *continuous line*—plant output) for 500 (S_{1T}; S_{2T}) and for 30 (S_{1T}; S_{2T}) iterations of L-M RTNN learning

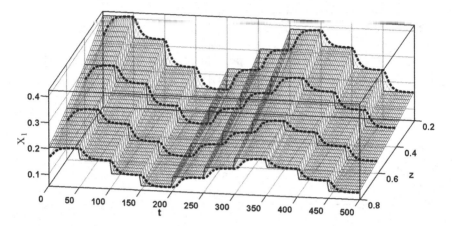

Fig. 1.31 Graphics of the 3D view of X_1 space/time approximation and indirect decentralized HFNMM I-term control in four collocation points of the fixed bed

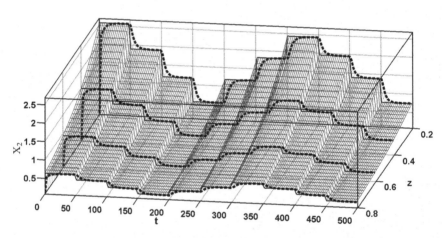

Fig. 1.32 Graphics of the 3D view of X_2 space/time approximation and indirect (SMC) decentralized HFNMM I-term control in four collocation points of the fixed bed

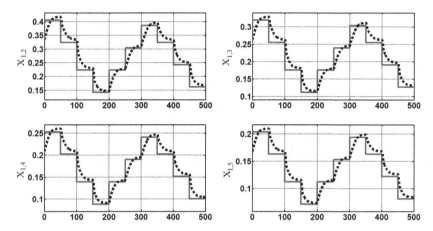

Fig. 1.33 Graphical simulation results of the indirect decentralized HFNMM proportional control (without I-term) of X_1 (acidogenic bacteria in the fixed bed) (*dotted line*—plant output, *continuous*-reference) in four collocation points (0.2H, 0.4H, 0.6H, 0.8H) for 500 L-M iterations ($X_{1,2}$; $X_{1,3}$; $X_{1,4}$; $X_{1,5}$)

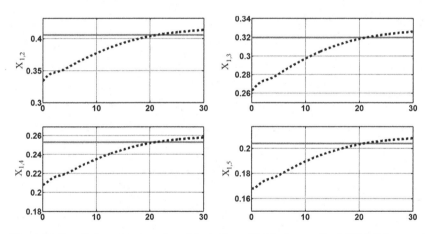

Fig. 1.34 Detailed graphical results of the indirect (SMC) decentralized HFNMM proportional control (without I-term) of X_1 (acidogenic bacteria in the fixed bed) (*dotted line*—plant output, *continuous*-reference) in 4 collocation points (0.2H, 0.4H, 0.6H, 0.8H) for the first 30 iterations ($X_{1,2}$; $X_{1,3}$; $X_{1,4}$; $X_{1,5}$)

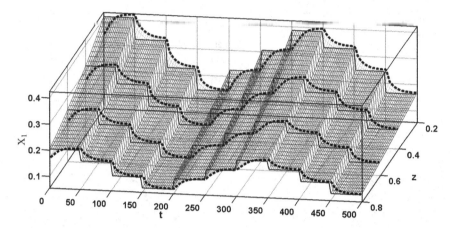

Fig. 1.35 Graphics of the 3D view of X_1 space/time approximation and indirect decentralized HFNMM proportional control (without I-term) in four collocation points of the fixed bed

Table 1.6 MSE of the indirect decentralized HFNMM I-term control of the bioprocess plant

Collocation point	X_1	X_2	S_1/S_{1T}	S_2/S_{2T}
$z = 0.2$	0.0004	0.0143	0.0449	0.2755
$z = 0.4$	0.0003	0.0047	0.0293	0.1019
$z = 0.6$	0.0002	0.0018	0.0192	0.0440
$z = 0.8$	0.0001	0.0009	0.0130	0.0235
Recirculation tank			0.0130	0.0263

Table 1.7 MSE of the indirect decentralized HFNMM proportional control (without I-term) of the bioprocess plant

Collocation point	X_1	X_2	S_1/S_{1T}	S_2/S_{2T}
$z = 0.2$	0.0005	0.0326	0.0457	1.5905
$z = 0.4$	0.0003	0.0095	0.0298	0.4036
$z = 0.6$	0.0002	0.0032	0.0195	0.1260
$z = 0.8$	0.0001	0.0013	0.0133	0.0516
Recirculation tank			0.0132	0.0450

1.7 Conclusion

The chapter performed decentralized recurrent fuzzy-neural identification, direct and indirect control of an anaerobic digestion wastewater treatment bioprocess, composed by a fixed bed and a recirculation tank, represented a DPS. The simplification of the PDE process model by ODE is realized using the orthogonal collocation method in three collocation points (plus the recirculation tank)

represented centers of membership functions of the space fuzzified output variables. The FNMM identifier used a second order Levenberg-Marquardt learning algorithm so to estimate parameters and states of the distributed parameters bioprocess plant. The obtained from the FNMMI state and parameter information is used by a HFNMM direct and indirect (sliding mode) control with I-term. The applied fuzzy-neural approach to that DPS decentralized direct and indirect identification and control exhibited a good convergence and precise reference tracking, which could be observed in the MSE numerical results given on Tables 1.4 and 1.6 (0.1965 vs. 0.2755 in the worse case) giving slight priority of the DANC over the IANC. The graphical and numerical results show that the I-term control eliminates constant noise and the proportional control could not.

Acknowledgments The Ph.D. student Eloy Echeverria Saldierna is thankful to CONACYT, Mexico for the scholarship received during his studies in the Department of Automatic Control, CINVESTAV-IPN, Mexico City, MEXICO.

Appendix 1: Detailed Derivation of the Recursive Levenberg-Marquardt Optimal Learning Algorithm for the RTNN

First of all we shall describe the optimal off-line learning method of Newton, then we shall modify it passing through the Gauss-Newton method and finally we shall simplify it so to obtain the off line Levenberg-Marquardt learning which finally will be transformed to recursive form (see [34] for more details).

The quadratic cost performance index under consideration is denoted by $J_k(W)$, where W is the RTNN vector of weights with dimension N_w subject of iterative learning during the cost minimization. Let us assume that the performance index is an analytic function so all its derivatives exist.

Let us to expand $J_k(W)$ around the optimal point of $W(k)$ which yields:

$$
\begin{aligned}
J_k(W) \approx & J_k(W(k)) + \nabla J_k^T(W(k))[W - W(k)] \\
& + \frac{1}{2}[W - W(k)]^T \nabla^2 J_k(W(k))[W - W(k)]
\end{aligned}
\tag{A.1}
$$

where $\nabla J(W)$ is the gradient of $J(W)$ with respect to the weight vector W:

$$
\nabla J(W) = \begin{bmatrix} \frac{\partial}{\partial W_1} J(W) \\ \frac{\partial}{\partial W_2} J(W) \\ \vdots \\ \frac{\partial}{\partial W_{N_W}} J(W) \end{bmatrix}
\tag{A.2}
$$

and $\nabla^2 J(W)$ is the Hessian matrix defined as:

$$\nabla^2 J(W) = \begin{bmatrix} \frac{\partial^2}{\partial w_1^2} J(W) & \frac{\partial^2}{\partial w_1 \partial w_2} J(W) & \cdots & \frac{\partial^2}{\partial w_1 \partial w_{N_w}} J(W) \\ \frac{\partial^2}{\partial w_2 \partial w_1} J(W) & \frac{\partial^2}{\partial w_2^2} J(W) & \cdots & \frac{\partial^2}{\partial w_2 \partial w_{N_w}} J(W) \\ \vdots & \vdots & \ddots & \vdots \\ \frac{\partial^2}{\partial w_{N_w} \partial w_1} J(W) & \frac{\partial^2}{\partial w_{N_w} \partial w_2} J(W) & \cdots & \frac{\partial^2}{\partial w_{N_w}^2} J(W) \end{bmatrix} \tag{A.3}$$

Taking the gradient of the Eq. (A.1) with respect to W and equating it to zero, we obtained:

$$\nabla J_k(W(k)) + \nabla^2 J_k(W(k))[W - W(k)] = 0 \tag{A.4}$$

Deriving (A.4) for W, we have:

$$W = W(k) - \left(\nabla^2 J_k(W(k))\right)^{-1} \nabla J_k(W(k)) \tag{A.5}$$

Finally, we obtain the Newton's learning algorithm as:

$$W(k+1) = W(k) - \left(\nabla^2 J_k(W(k))\right)^{-1} \nabla J_k(W(k)) \tag{A.6}$$

where $W(k+1)$ is the weight vector minimizing $J_k(W)$ in the instant k; i.e. $J_k(W)|_{W=W(k+1)}$ is min.

Let us suppose that $W(k)$ is the weight vector that minimize $J_{k-1}(W)$ in the instant $k-1$, then:

$$\nabla J_{k-1}(W)|_{W=W(k)} = \nabla J_{k-1}(W(k)) = 0 \tag{A.7}$$

The performance index is defined as:

$$J_k(W) = \frac{1}{2} \sum_{q=1}^{k} \alpha^{k-q} E_q^T(W) E_q(W) \tag{A.8}$$

$$J_k(W) = \frac{1}{2} \sum_{q=1}^{k} \left(\alpha^{k-q} \sum_{j=1}^{L} e_{j,q}^2(W) \right) \tag{A.9}$$

where: $0 < \alpha \leq 1$ is a forgetting factor, q is the instant of the corresponding error vector, E_q represented the qth error vector, $e_{j,q}$ is the jth element of E_q, k es is the final instant of the performance index. The ith element of the gradient is:

$$[\nabla J_k(W)]_i = \frac{\partial J_k(W)}{\partial w_i} = \sum_{q=1}^{k} \left(\alpha^{k-q} \sum_{j=1}^{L} e_{j,q}(W) \frac{\partial e_{j,q}(W)}{\partial w_i} \right) \tag{A.10}$$

$$[\nabla J_k(\mathbf{W})]_i = \sum_{q=1}^{k} \left(\alpha^{k-q} \sum_{j=1}^{L} e_{j,q}(\mathbf{W}) \frac{\partial (r_{j,q} - y_{j,q}(\mathbf{W}))}{\partial w_i} \right)$$

$$= -\sum_{q=1}^{k} \left(\alpha^{k-q} \sum_{j=1}^{L} e_{j,q}(\mathbf{W}) \frac{\partial y_{j,q}(\mathbf{W})}{\partial w_i} \right) \tag{A.11}$$

The matricial form of the performance index gradient is:

$$\nabla J_k(\mathbf{W}) = -\sum_{q=1}^{k} \alpha^{k-q} \mathbf{J}_{\mathbf{Y}_q}^T(\mathbf{W}) \mathbf{E}_q(\mathbf{W}) \tag{A.12}$$

where the Jacobean matrix of $\mathbf{Y}_q$ in the instant q with dimension $L \times N_w$ is:

$$\mathbf{J}_{\mathbf{Y}_q}(\mathbf{W}) = \begin{bmatrix} \frac{\partial y_{1,q}}{\partial w_1} & \frac{\partial y_{1,q}}{\partial w_2} & \cdots & \frac{\partial y_{1,q}}{\partial w_{N_w}} \\ \frac{\partial y_{2,q}}{\partial w_1} & \frac{\partial y_{2,q}}{\partial w_2} & \cdots & \frac{\partial y_{2,q}}{\partial w_{N_w}} \\ \vdots & \vdots & \ddots & \vdots \\ \frac{\partial y_{L,q}}{\partial w_1} & \frac{\partial y_{L,q}}{\partial w_2} & \cdots & \frac{\partial y_{L,q}}{\partial w_{N_w}} \end{bmatrix} \tag{A.13}$$

The gradient could be written in the following form:

$$\nabla J_k(\mathbf{W}) = \alpha \nabla J_{k-1}(\mathbf{W}) - \mathbf{J}_{\mathbf{Y}_k}^T(\mathbf{W}) \mathbf{E}_k(\mathbf{W}) \tag{A.14}$$

Then, the h, ith element of the Hessian matrix could be written as:

$$[\nabla^2 J_k(\mathbf{W})]_{h,i} = \frac{\partial^2 J_k(\mathbf{W})}{\partial w_h \partial w_i}$$

$$= \sum_{q=1}^{k} \alpha^{k-q} \sum_{j=1}^{L} \left(\frac{\partial e_{j,q}(\mathbf{W})}{\partial w_h} \frac{\partial e_{j,q}(\mathbf{W})}{\partial w_i} + e_{j,q}(\mathbf{W}) \frac{\partial^2 e_{j,q}(\mathbf{W})}{\partial w_h \partial w_i} \right) \tag{A.15}$$

$$[\nabla^2 J_k(\mathbf{W})]_{h,i} = \sum_{q=1}^{k} \alpha^{k-q} \sum_{j=1}^{L} \left(\frac{\partial e_{j,q}(\mathbf{W})}{\partial w_h} \frac{\partial e_{j,q}(\mathbf{W})}{\partial w_i} + e_{j,q}(\mathbf{W}) \frac{\partial^2 e_{j,q}(\mathbf{W})}{\partial w_h \partial w_i} \right) \tag{A.16}$$

$$\nabla^2 J_k(\mathbf{W}) = \sum_{q=1}^{k} \alpha^{k-q} \left(\mathbf{J}_{\mathbf{Y}_q}^T(\mathbf{W}) \mathbf{J}_{\mathbf{Y}_q}(\mathbf{W}) + \sum_{j=1}^{L} e_{j,q}(\mathbf{W}) \nabla^2 e_{j,q}(\mathbf{W}) \right) \tag{A.17}$$

Equating:

$$\sum_{j=1}^{L} e_{j,q}(\mathbf{W}) \nabla^2 e_{j,q}(\mathbf{W}) \approx 0 \tag{A.18}$$

we could obtain directly the Gauss-Newton method of optimal learning.

The Eq. (A.17) is reduced to:

$$\nabla^2 J_k(\mathbf{W}) = \sum_{q=1}^{k} \alpha^{k-q} J_{Y_q}^T(\mathbf{W}) J_{Y_q}(\mathbf{W}) \tag{A.19}$$

and it could also be written as:

$$\nabla^2 J_k(\mathbf{W}) = \alpha \nabla^2 J_{k-1}(\mathbf{W}) + J_{Y_k}^T(\mathbf{W}) J_{Y_k}(\mathbf{W}) \tag{A.20}$$

Then (A.14) and (A.20) solved for $\mathbf{W} = \mathbf{W}(k)$ are transformed to:

$$\nabla J_k(\mathbf{W}(k)) = \alpha \nabla J_{k-1}(\mathbf{W}(k)) - J_{Y_k}^T(\mathbf{W}(k)) E_k(\mathbf{W}(k)) \tag{A.21}$$

$$\nabla^2 J_k(\mathbf{W}(k)) = \alpha \nabla^2 J_{k-1}(\mathbf{W}(k)) + J_{Y_k}^T(\mathbf{W}(k)) J_{Y_k}(\mathbf{W}(k)) \tag{A.22}$$

According to (A.7), the Eq. (A.21) is reduced to:

$$\nabla J_k(\mathbf{W}(k)) = -J_{Y_k}^T(\mathbf{W}(k)) E_k(\mathbf{W}(k)) \tag{A.23}$$

Let us define:

$$H(k) = \nabla^2 J_k(\mathbf{W}(k)) \tag{A.24}$$

Then we could write the Eq. (A.22) in the following form:

$$H(k) = \alpha H(k-1) + J_{Y_k}^T(\mathbf{W}(k)) J_{Y_k}(\mathbf{W}(k)) \tag{A.25}$$

Finally for the learning algorithm (A.6), we could obtain:

$$\mathbf{W}(k+1) = \mathbf{W}(k) + \left(\nabla^2 J_k(\mathbf{W}(k))\right)^{-1} J_{Y_k}^T(\mathbf{W}(k)) E_k(\mathbf{W}(k)) \tag{A.26}$$

$$\mathbf{W}(k+1) = \mathbf{W}(k) + H^{-1}(k) J_{Y_k}^T(\mathbf{W}(k)) E_k(\mathbf{W}(k)) \tag{A.27}$$

The Eq. (A.27) corresponds to the Gauss-Newton learning method where the considered Hessian matrix is an approximation to the real one.

Let us now to come back to Eq. (A.19).

$$H(k) = \nabla^2 J_k(\mathbf{W}) = \sum_{q=1}^{k} \alpha^{k-q} J_{Y_q}^T(\mathbf{W}) J_{Y_q}(\mathbf{W}) \tag{A.28}$$

Here we observe that the product $J^T J$ could be nonsingular which require to perform the following modification of the Hessian matrix:

$$H(k) = \nabla^2 J_k(W) = \sum_{q=1}^{k} \alpha^{k-q} \left(J_{Y_q}^T(W) J_{Y_q}(W) + \rho I \right) \qquad (A.29)$$

The Hessian matrix could be written also in the form:

$$H(k) = \alpha H(k-1) + J_{Y_k}^T(W) J_{Y_k}(W) + \rho I \qquad (A.30)$$

where ρ is a small constant (generally ρ is chosen between 10^{-2} and 10^{-4}).

This modification of the Hessian matrix is essential for the optimal learning method of Levenberg-Marquardt. The computation of the Hessian matrix inverse could be done using the matrix inversion lemma which requires the following modification:

$$H(k) = \alpha H(k-1) + J_{Y_k}^T(W) J_{Y_k}(W) + \rho I_{N_w} \qquad (A.31)$$

where I_{N_w} is a $N_w \times N_w$ zero matrix except one element (with value 1) corresponding to position $i = (k \bmod N_w) + 1$. It could be seen that after N_w iterations, the Eq. (A.31) become equal to Eq. (A.30), i.e.:

$$\sum_{n=k+1}^{k+N_w} \rho I_{N_w}(n) = \rho I \qquad (A.32)$$

Then the Eq. (A.31) is transformed to:

$$H(k) = \alpha H(k-1) + \Omega^T(k) \Lambda^{-1}(k) \Omega(k) \qquad (A.33)$$

where:

$$\Omega(k) = \begin{bmatrix} & & & J_{Y_k}(W) & & \\ 0 & \cdots & 0 & 1 & 0 & \cdots & 0 \end{bmatrix} \qquad (A.34)$$

$$\Lambda(k)^{-1} = \begin{bmatrix} I & 0 \\ 0 & \rho \end{bmatrix} \qquad (A.35)$$

Then it is easy to apply the matrix inversion lemma, which constitute to the following equation (where the matrices A, B, C and D have compatible dimensions and the product BCD, and the sum A + BCD exists):

$$[A + BCD]^{-1} = A^{-1} - A^{-1}B \left[DA^{-1}B + C^{-1} \right]^{-1} DA^{-1} \qquad (A.36)$$

Let us to apply the following substitutions:

$$A = \alpha H(k-1); B = \Omega^T(k); C = \Lambda^{-1}(k); D = \Omega(k)$$

The inverse of the Hessian matrix $H(k)$ could be computed using the expression:

$$H^{-1}(k) = \left[\alpha H(k-1) + \Omega^T(k)\Lambda^{-1}(k)\Omega(k)\right]^{-1} = \alpha^{-1}H^{-1}(k-1)$$
$$- \alpha^{-1}H^{-1}(k-1)\Omega^T(k)\left[\Omega(k)\alpha^{-1}H^{-1}(k-1)\Omega^T(k) + \Lambda(k)\right]^{-1}$$
$$\Omega(k)\alpha^{-1}H^{-1}(k-1)$$

$$(A.37)$$

$$H^{-1}(k) = \alpha^{-1}\Big\{H^{-1}(k-1)$$
$$- H^{-1}(k-1)\Omega^T(k)\left[\Omega(k)H^{-1}(k-1)\Omega^T(k) + \alpha\Lambda(k)\right]^{-1} \quad (A.38)$$
$$\Omega(k)H^{-1}(k-1)\Big\}$$

Let us denote

$$P(k) = H^{-1}(k)$$

and substitute it in the Eq. (A.38), we obtained:

$$P(k) = \alpha^{-1}\{P(k-1) - P(k-1)\Omega^T(k)S^{-1}(k)\Omega(k)P(k-1)\} \quad (A.39)$$

where:

$$S(k) = \alpha\Lambda(k) + \Omega(k)P(k-1)\Omega^T(k) \quad (A.40)$$

Finally, the learning algorithm for $\mathbf{w}$ is obtained as:

$$W(k+1) = W(k) + P(k)J_{Y_k}^T(W(k))E_k(W(k)) \quad (A.41)$$

where W is a $N_w \times 1$ vector formed of all RTNN weights ($N_w = L \times N + N + N \times M$).

Using the RTNN topology the weight vector has the following form:

$$W(k) = [c_{1,1} \quad \cdots \quad c_{L,N} \quad a_{1,1} \quad a_{2,2} \quad \cdots \quad a_{N,N} \quad b_{1,1} \quad \cdots \quad b_{N,M}]^T \quad (A.42)$$

and the Jacobean matrix with dimension $L \times N_w$. is formed as:

$$J_{Y_k}(W(k)) = [J_{Y_k}(C(k)) \quad J_{Y_k}(A(k)) \quad J_{Y_k}(B(k))] \quad (A.43)$$

The components of the Jacobean matrix could be obtained applying the diagrammatic method [31]. Using the notation of part 2.2 for (A.43), we could write:

$$DY[W(k)] = \left[DY\big(C_{ij}(k)\big), DY\big(A_{ij}(k)\big), DY\big(B_{ij}(k)\big)\right].$$

Appendix 2

Table A1 Abbreviations used in the chapter

Abbreviation	Term
ANN	Artificial neural networks
BP	Backpropagation
CI	Computational intelligence
DPS	Distributed parameter systems
FB-FF	Feedback-feedforward
FFNN	Feedforward neural networks
FNMM	Fuzzy-neural multi-model
FNMMI	Fuzzy-neural multi-model identifier
FNMMC	Fuzzy-neural multi-model controller
FNMMFBC	Fuzzy-neural multi-model feedback controller
FRBIS	Fuzzy rule-based inference system
FS	Fuzzy systems
HFNMM	Hierarchical fuzzy-neural multi-model
IANC	Indirect adaptive neural control
I-term	Integral term
L-M	Levenberg-Marquardt
LLC	Lower level of control
MSE	Means squared error
NN	Neural network
OCM	Orthogonal collocation method
ODE	Ordinary differential equations
PDE	Partial differential equations
RNNM	Recurrent neural network model
RTNN	Recurrent trainable neural network
RNN	Recurrent neural networks
SM	Sliding mode
SMC	Sliding mode control
T-S	Takagi-Sugeno
ULC	Upper level of control

References

1. Haykin, S.: Neural Networks: A Comprehensive Foundation, 2nd edn, Section 2.13, 84–89; Section 4.13, pp. 208–213. Prentice-Hall, Upper Saddle River (1999)
2. Narendra, K.S., Parthasarathy, K.: Identification and control of dynamical systems using neural networks. IEEE Trans. Neural Networks **1**(1), 4–27 (1990)
3. Chen, S., Billings, S.A.: Neural networks for nonlinear dynamics system modelling and identification. Int. J. Control **56**(2), 319–346 (1992)
4. Hunt, K.J., Sbarbaro, D., Zbikowski, R., Gawthrop, P.J.: Neural network for control systems (A survey). Automatica **28**, 1083–1112 (1992)

5. Miller III, W.T., Sutton, R.S., Werbos, P.J.: Neural Networks for Control. MIT Press, London (1992)
6. Pao, S.A., Phillips, S.M., Sobajic, D.J.: Neural net computing and intelligent control systems. Int. J. Control **56**(3), 263–289 (1992). (Special issue on Intelligent Control)
7. Su, H.-T., McAvoy, T.J., Werbos, P.: Long-term predictions of chemical processes using recurrent neural networks: a parallel training approach. Ind. Eng. Chem. Res. **31**(5), 1338–1352 (1992)
8. Boskovic, J.D., Narendra, K.S.: Comparison of linear, nonlinear and neural-network-based adaptive controllers for a class of fed-batch fermentation processes. Automatica **31**, 817–840 (1995)
9. Omatu, S., Khalil, M., Yusof, R.: Neuro-Control and Its Applications. Springer, London (1995)
10. Baruch, I.S., Garrido, R.: A direct adaptive neural control scheme with integral terms. Int. J. Intell. Syst. **20**(2), 213–224 (2005). ISSN 0884-8173. (Special issue on Soft Computing for Modelling, Simulation and Control of Nonlinear Dynamical Systems, Castillo, O., Melin, P. guest ed. Wiley Inter-Science)
11. Bulsari, A., Palosaari, S.: Application of neural networks for system identification of an adsorption column. Neural Comput. Appl. **1**, 160–165 (1993)
12. Deng, H., Li, H.X.: Hybrid intelligence based modelling for nonlinear distributed parameter process with applications to the curing process. IEEE Trans. Syst. Man Cybern. **4**, 3506–3511 (2003)
13. Deng, H., Li, H.X.: Spectral-approximation-based intelligent modelling for distributed thermal processes. IEEE Trans. Control Syst. Technol **13**, 686–700 (2005)
14. Gonzalez-Garcia, R., Rico-Martinez, R., Kevrekidis, I.: Identification of distributed parameter systems: a neural net based approach. Comput. Chem. Eng. **22**(4-supl. 1), 965–968 (1998)
15. Padhi, R., Balakrishnan, S., Randolph, T.: Adaptive critic based optimal neuro- control synthesis for distributed parameter systems. Automatica **37**, 1223–1234 (2001)
16. Padhi, R., Balakrishnan, S.: Proper orthogonal decomposition based optimal neuro-control synthesis of a chemical reactor process using approximate dynamic programming. Neural Networks **16**, 719–728 (2003)
17. Pietil, S., Koivo, H.N.: Centralized and decentralized neural network models for distributed parameter systems. In: Proceedings of the Symposium on Control, Optimization and Supervision, CESA'96, IMACS Multiconference on Computational Engineering in Systems Applications, Lille, France, pp. 1043–1048 (1996)
18. Lin, C.T., Lee, C.S.G.: Neural Fuzzy Systems: A Neuro—Fuzzy Synergism to Intelligent Systems. Prentice Hall, Englewood Cliffs (1996)
19. Babuska, R.: Fuzzy Modeling for Control. Kluwer, Norwell (1998)
20. Baruch, I., Beltran-Lopez, R., Olivares-Guzman, J.L., Flores, J.M.: A fuzzy-neural multi-model for nonlinear systems identification and control. Fuzzy Sets Syst. **159**, 2650–2667 (2008)
21. Takagi, T., Sugeno, M.: Fuzzy identification of systems and its applications to modelling and control. IEEE Trans. Syst. Man Cybern. **15**, 116–132 (1985)
22. Teixeira, M., Zak, S.: Stabilizing controller design for uncertain nonlinear systems, using fuzzy models. IEEE Trans. Syst. Man Cybern. **7**, 133–142 (1999)
23. Mastorocostas, P.A., Theocharis, J.B.: A recurrent fuzzy-neural model for dynamic system identification. IEEE Trans. Syst. Man Cybern. B Cybern. **32**, 176–190 (2002)
24. Mastorocostas, P.A., Theocharis, J.B.: An orthogonal least-squares method for recurrent fuzzy-neural modeling. Fuzzy Sets Syst. **140**(2), 285–300 (2003)
25. Galvan-Guerra, R., Baruch, I.S.: Anaerobic digestion process identification using recurrent neural multi-model. In: Gelbukh, A., Kuri-Morales, A.F. (eds.) Sixth Mexican International Conference on Artificial Intelligence, 4–10 Nov 2007, Aguascalientes, Mexico, Special Session, Revised Papers, CPS, pp. 319–329. IEEE Computer Society, Los Alamos. ISBN 978-0-7695-3124-3 (2008)

26. Baruch, I., Olivares-Guzman, J.L., Mariaca-Gaspar, C.R., Galvan-Guerra, R.: A sliding mode control using fuzzy-neural hierarchical multi-model identifier. In: Castillo, O., Melin, P., Ross, O.M., Cruz, R.S., Pedrycz, W., Kacprzyk, J. (eds.) Theoretical Advances and Applications of Fuzzy Logic and Soft Computing, ASC, vol. 42, pp. 762–771. Springer, Berlin (2007)

27. Baruch, I., Olivares-Guzman, J.L., Mariaca-Gaspar, C.R., Galvan-Guerra, R.: A fuzzy-neural hierarchical multi-model for systems identification and direct adaptive control. In: Melin, P., Castillo, O., Ramirez, E.G., Kacprzyk, J., Pedrycz, W. (eds.) Analysis and Design of Intelligent Systems Using Soft Computing Techniques, ASC, vol. 41, pp. 163–172. Springer, Berlin (2007)

28. Aguilar-Garnica, F., Alcaraz-Gonzalez, V., Gonzalez-Alvarez, V.: Interval observer design for an anaerobic digestion process described by a distributed parameter model. In: Proceedings of the 2nd International Meeting on Environmental Biotechnology and Engineering (2IMEBE), CINVESTAV-IPN, Mexico City, paper 117 (2006), pp. 1–16

29. Bialecki, B., Fairweather, G.: Orthogonal spline collocation methods for partial differential equations. J. Comput. Appl. Math. **128**, 55–82 (2001)

30. Baruch, I.S., Mariaca-Gaspar, C.R.: A Levenberg-Marquardt learning applied for recurrent neural identification and control of a wastewater treatment bioprocess. Int. J. Intell.Syst. **24**, 1094–1114 (2009). ISSN 0884-8173

31. Wan, E., Beaufays, F.: Diagrammatic method for deriving and relating temporal neural network algorithms. Neural Comput. **8**, 182–201 (1996)

32. Nava, F., Baruch, I.S., Poznyak, A., Nenkova, B.: Stability proofs of advanced recurrent neural networks topology and learning. Comptes Rendus, **57**(1), 27–32 (2004). ISSN 0861-1459. (Proceedings of the Bulgarian Academy of Sciences)

33. Baruch, I.S., Mariaca-Gaspar, C.R., Barrera-Cortes, J.: Recurrent neural network identification and adaptive neural control of hydrocarbon biodegradation processes. In: Hu, X., Balasubramaniam, P. (eds.) Recurrent Neural Networks, Chapter 4, pp. 61–88. I-Tech Education and Publishing KG, Vienna (2008). ISBN 978-953-7619-08-4

34. Ngia, L.S., Sjöberg, J.: Efficient training of neural nets for nonlinear adaptive filtering using a recursive Levenberg-Marquardt algorithm. IEEE Trans. Signal Process. **48**, 1915–1927 (2000)

Chapter 2
Error Tolerant Predictive Control Based on Recurrent Neural Models

Petia Georgieva and Sebastião Feyo de Azevedo

Abstract This chapter is focused on developing a feasible model predictive control (MPC) based on time dependent recurrent neural network (NN) models. A modification of the classical regression neural models is proposed suitable for prediction purposes. In order to reduce the computational complexity and to improve the prediction ability of the neural model, optimization of the NN structure (lag space selection, number of hidden nodes), pruning techniques and identification strategies are discussed. Furthermore a computationally efficient modification of the general nonlinear MPC is proposed termed Error Tolerant MPC (ETMPC). The NN model is imbedded into the structure of the ETMPC and extensively tested on a dynamic simulator of an industrial crystalizer. The results demonstrate that the NN-based ETMPC relaxes the computational burden without losing closed loop performance and can complement other solutions for feasible industrial real time control.

2.1 Introduction

The concept of Model-based Predictive Control (MPC) was introduced in the eighties [1]. It does not refer to a particular control method, instead it corresponds to a general control framework [2]. MPC is known to be the most successful advanced control approach in practical applications, representing a true alternative to the classical Proportional Integral Derivative (PID) control. The main reasons

P. Georgieva (✉)
Signal Processing Lab, Department of Electronics Telecommunications and Informatics (DETI), Institute of Electronics Engineering and Telematics of Aveiro (IEETA), University of Aveiro, Aveiro, Portugal
e-mail: petia@ua.pt

S. F. de Azevedo
Faculdade de Engenharia de Universidade do Porto, Porto, Portugal

V. E. Balas et al. (eds.), *Innovations in Intelligent Machines-5*,
Studies in Computational Intelligence 561, DOI: 10.1007/978-3-662-43370-6_2,
© Springer-Verlag Berlin Heidelberg 2014

for this are its potential i) to take directly into account the process input/output constrains, ii) to consider multiple process objectives and iii) to control processes with nonlinear time varying dynamics. The main difference between the existing MPC configurations is in the model used to predict the future behavior of the process or the implemented optimization procedure. The first industrial applications of MPC were based on linear models [3], later on successful applications of nonlinear MPC (NMPC) [4] were also reported. Despite the recognized progress of NMPC, its on-line implementation with predictions running on a large number of nonlinear differential and algebraic equations (DAE), i.e. the first principles process model, is a huge challenge. Such predictions may lead to feasibility problems for processes with fast nonlinear dynamics (as for example chemical crystallization/precipitation processes) [5] or can stuck into numerical problems as stiffness or ill-conditioning. Moreover, in many cases development of first principles models based on physical laws is difficult and time consuming [6].

Several suggestions have been made to deal with these problems ranging from simple extension of Dynamic Matrix Control (DMC) based on successive linearization of the nonlinear model to more elaborate techniques involving discretization of the model followed by solution via non-linear programming [7]. These solutions are usually computationally very intensive, assuming unlimited computing resources, which is only valid for high value products and industries with state of the art control equipment.

MPC algorithms based on artificial Neural Network (NN) models are a promising alternative that addresses also the above problems. Among various NN structures, one-step ahead predictors where the neural model is trained non-recurrently is the most typical option e.g., [8, 9]. However, if the neural models are used for long range prediction (usually the case with MPC), the prediction error will be propagated. Another option is to use a multi-model [10, 11]. For each sampling instant within the prediction horizon an independent submodel is used, thus the prediction error is not propagated. A third option is to use specialised recurrent training algorithms for neural models [7, 12, 13], but they are significantly more computationally demanding in comparison with one-step ahead predictor training, and the obtained models may be sensitive to noise.

The contribution of this chapter is twofold. First, it introduces a recurrent multistep ahead predictive neural model where the past model predictions are substituted by the process measurements and thus the prediction error is not propagated. Secondly, in order to further improve the efficiency and reduce the complexity of the control system a modification of the general NMPC is also proposed, where the NMPC is executed only when the tracking error is outside a pre-specified bound. Once the error converges towards the tolerance zone, the NMPC is switched off and the control action is kept constant. The combination of the proposed neural model and the introduced error tolerance (ET) in the optimization represents a promising compromise between process performance and computational complexity and can complement other suggestions in the literature for feasible industrial real time control. This modification is termed NN-based Error Tolerant Model Predictive Controller (NN ETMPC).

2.2 Model Predictive Control Algorithms

Although individually different in form, the underlying idea of all MPC schemes can be summarized as follows [14]: i) A dynamic model and on-line measurements are used to predict the future process behavior; ii) On the basis of the process output predictions over a prediction horizon (H_p), optimization is performed to find a sequence of manipulated input moves over a control horizon (H_c) that minimizes a chosen cost function while satisfying all given constrains; iii) Only the first of the calculated input sequence is implemented and the whole optimization is repeated at the next sampling time (see Fig. 2.1).

The on-line NMPC implementation has often a discrete form, where the exact optimization is substituted by a discrete approximation [15]. The continuous time $t \in [t_0, t_f]$ is divided into equally spaced time intervals, with discrete time steps $t_k = t_0 + k\Delta t$ and $k = 0, 1, ..., N$. At each consecutive sampling instant (k), a set of future $(k + p)$ control increments $\Delta u(k + p/k) = u(k + p/k) - u(k + p - 1/k)$ is calculated

$$\Delta \mathbf{u}(k) = [\Delta u(k/k), \, \Delta u(k + 1/k), ... \Delta u(k + H_c - 1/k)] \tag{2.1}$$

The following quadratic cost function is typically used:

$$\min_{\Delta u(k/k), \, \Delta u(k+1/k)......\Delta u(k+H_c-1/k)} J = \lambda_1 \sum_{p=1}^{H_p} (e(k + p/k))^2 + \lambda_2 \sum_{p=1}^{H_c} (\Delta u(k + p/k))^2 \tag{2.2}$$

where $e(k + p/k) = ref(k + p/k) - \hat{y}(k + p/k)$. Note that the first term in (2.2) is related with the objective to minimize the deviation of the predicted values of the output $\hat{y}(k + p/k)$ from the respective reference $ref\,(k + p/k)$ over the prediction horizon. The second term penalizes excessive control increments. The prediction horizon (H_p) is the number of future time steps over which the prediction errors are minimized and the control horizon (H_c) is the number of future time steps over which the control increments are minimized, $H_p \geq H_c$. It is assumed that the control increments for sampling instants after the control horizon are zero ($\Delta u(k + p/k) = 0, p \geq H_c$). Parameters $\{\lambda_1, \lambda_2\} \in \Re \geq 0$ determine the contribution of each term of (2.2) and consider also the problem of different numerical ranges.

The cost function (2.2) has to be minimized subject to the following constraints:

$$u_{\min} \leq u(k + p/k) \leq u_{\max}, \quad p = 0, 1, ... H_c - 1 \tag{2.3}$$

$$\Delta u_{\min} \leq \Delta u(k + p/k) \leq \Delta u_{\max}, \quad p = 0, 1, ... H_c - 1 \tag{2.4}$$

$$y_{\min} \leq \hat{y}(k + p/k) \leq y_{\max}, \quad p = 1, 2, ... H_p \tag{2.5}$$

$u_{\min}$ and $u_{\max}$ are the lower and the upper bounds of the manipulated inputs, while $\Delta u_{\min}$ and $\Delta u_{\max}$ are the limits of the manipulated input increments.

Fig. 2.1 Model based
predictive control (MPC)
scheme

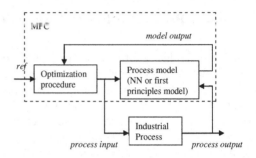

Constraints (2.3–2.4) are usually related with actuator saturation or rate-of-change restrictions, whereas constraints (2.5) are associated with operational limits (y_{min}, y_{max}) such as equipment specifications and safety considerations. Only the first element of the determined sequence (2.1) is actually applied to the process

$$u(k) = \Delta u(k/k) + u(k-1) \tag{2.6}$$

The general prediction equation for $p = 1, 2, \ldots H_p$ is

$$\hat{y}(k+p/k) = y_m(k+p/k) + dist(k) \tag{2.7}$$

where quantities $y_m(k+p/k)$ are calculated from a dynamical model of the process. The unmeasured disturbance $dist(k)$ is estimated from

$$dist(k) = y(k) - y_m(k/k-1) \tag{2.8}$$

where $y(k)$ is measured from the process and $y_m(k/k-1)$ is calculated from the dynamical model. $dist(k)$ is assumed to be constant over the prediction horizon $dist(k+p/k) = dist(k)$, $i = 1, \ldots H_c$.

MPC algorithms use an explicit dynamical model in order to predict the future process behavior (Fig. 2.1). Therefore, the main issue to address is the choice of the process model structure since it affects the performance and accuracy of the control action. Models based on physical laws (first principles models) are usually very precise, however not suitable for on-line control since they are complicated and may lead to numerical problems. Linear models are approximations and a good choice when the process nonlinearity is mild. However, when the process is substantially nonlinear, data-based models as fuzzy and neural network structures are gaining more popularity [10].

On Fig. 2.2 are summarized the most typical MPC algorithms. When the model is linear and there are no constrains affecting the process, an analytical optimal solution can be found in a closed form [2]. This is however an idealization, while in practice the linear MPC is solved numerically at each sampling instant, by the constrained quadratic programming (QP) optimization. When the model is nonlinear, then the optimization problem (2.2) is certainly not linear quadratic. It is generally a nonconvex and even multimodal one. For such problems there are no

Fig. 2.2 MPC algorithms

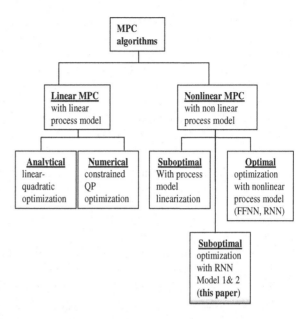

sufficiently fast and reliable numerical optimization procedures, i.e., procedures yielding always an optimal point and within predefined time limit—as is required in on-line control applications [13]. Therefore, many attempts have been made to construct simplified (and generally suboptimal) nonlinear MPC algorithms avoiding full on-line nonlinear optimization, first of all using model linearizations or multiple linear models (for example fuzzy structures) [11]. On the other hand, successful optimal MPC based on nonlinear optimization (MPC-NO) are mainly those applying NN techniques. However, the MPC-NO algorithm with NN is still computationally very demanding (nonconvex, local minima) and significantly more powerful controllers are required especially for shorter sampling periods.

How to reduce the computational complexity of the nonlinear MPC is therefore a pertinent problem. The contribution of this chapter is the combination of a relaxed cost function (suboptimal optimization) and the specific time dependent Recurrent Neural Network (RNN) structure in order to reduce the MPC computational burden.

2.3 Neural Dynamical Process Model

NNs became an established methodology for identification and control of nonlinear systems because they are universal approximators, have a relatively small number of parameters and a simple structure. The NN-based control can be approached in direct or indirect control framework. Direct neural control means that the controller itself is a NN, while in the indirect neural control scheme, first a NN is used to model the process to be controlled, and this model is then embedded in the control structure [16]. The first approach consists of a number of methods such as NN Inverse Control (known also as Adaptive Neural Control), NN Internal Model Control (IMC), NN feedback linearization, NN feedforward controller + conventional feedback controller [17]. The second approach is associated with Direct adaptive control and NN-based MPC [18].

The main MPC approaches based on NN dynamical models can be briefly summarized as:

(1) MPC in which the neural model is used directly, without any simplifications. The control action is computed by a nonlinear optimisation routine, e.g., [15, 16].
(2) MPC in which the neural model is linearised on-line. At each sampling instant the control action is computed by a QP routine, e.g., [10, 13].
(3) Approximate neural MPC in which the NN replaces the whole control algorithm. The network generates the control actions [17].
(4) There are some specific versions of the above approaches termed stable neural MPC [19], adaptive neural MPC [20], robust neural MPC [21].

All these algorithms use a one step ahead predictive structure for NN training, while the MPC implementation of the trained neural model is as a multi step ahead process predictor. This discrepancy may cause biased predictions and error accumulation. We propose here to use the same structure for training and MPC implementation by making changes of the classical regression neural structure.

The most typical NN regression models are inspired by the classical nonlinear function approximation techniques [22].

2.3.1 NN Final Impulse Response (NNFIR) Model

$$y_m(k) = g[u(k - \tau), \ldots u(k - \tau - n_u + 1)] \tag{2.9}$$

At each discrete moment k, the model output, $y_m(k)$, is a function only of n_u past process inputs $u(k - \tau) \ldots u(k - \tau - n_u + 1)$, τ is the discrete time delay, $\tau \leq n_u$.

2.3.2 NN Auto Regressive with eXogenous Input (NNARX) Model

$$y_m(k) = g\lfloor y(k - 1), \ldots y(k - n_y), u(k - \tau), \ldots u(k - \tau - n_u + 1)\rfloor \tag{2.10}$$

The network input consists of n_u past process (exogenous) inputs and n_y past process outputs $y(k - 1) \ldots y(k - n_y)$.

2.3.3 NN Auto Regressive Moving Average with eXogenous Input (NNARMAX) Model

$$y_m(k) = g\left[y(k-1), \ldots y(k-n_y), u(k-\tau), \ldots u(k-\tau-n_u+1), e(k-1), \ldots e(k-n_e)\right]$$
(2.11)

where $e(k) = y(k) - y_m(k)$.

The model output $y_m(k)$ is a function of n_u past process inputs, n_y past process outputs and n_e past errors between the process and the model output $e(k-1) \ldots e(k-n_e)$.

2.3.4 NN Output Error (NNOE) Model

$$y_m(k) = g\left\lfloor y_m(k-1), \ldots y_m(k-n_y), u(k-\tau), \ldots u(k-\tau-n_u+1)\right\rfloor \quad (2.12)$$

The network input consists of n_u past process inputs and n_y past model outputs $y_m(k-1) \ldots y_m(k-n_y)$.

NNFIR or NNARX models can be designed as a Feed Forward NN (FFNN), while to build a NNARMAX or NNOE model, a Recurrent Neural Network (RNN) with global feedback (network outputs are fed back as network inputs) is required. RNNs are more adequate to approximate process dynamics since they can encode the system real time memory and have usually a shorter lag space (n_y and n_u) [23].

The output of any of the four NN models (A–D) can be computed as

$$y_m(k) = a_0 + \sum_{j=1}^{H} a_j \varphi\left(z_j(k)\right)$$
(2.13)

where H is the number of hidden nodes and φ is the nonlinear activation function. $z_j(k)$ is the sum of inputs of the ith hidden node. For a standard RNN

$$z_j(k) = b_{j0} + \sum_{i=1}^{n_y} b_{ji} y(k-i) + \sum_{i=1}^{n_u} c_{ji} u(k-\tau-i+1)$$
(2.14)

$\{a_j, b_{j\text{-}}, c_{j\text{-}}\} \in \Re$ are the hidden-to-output layer weights and the input-to-hidden layer weights, respectively.

MPC requires multistep ahead predictions of the process outputs. If NNARX or NNARMAX are used as predictive models, according to (2.10) and (2.11), future process outputs (measurements) are necessary. For example, let $n_y = n_u = 3$,

$\tau = 1$, $H_p = 4$, $H_c = 2$. Assuming NNARX model structure, at each sampling moment k the following computations will be performed

$$
\begin{aligned}
y_m(k) &= g[y(k-1), y(k-2), y(k-3), u(k-1), u(k-2), u(k-3)] \\
y_m(k+1) &= g[y(k), y(k-1), y(k-2), u(k), u(k-1), u(k-2)] \\
y_m(k+2) &= g[y(k+1), y(k), y(k-1), u(k+1), u(k), u(k-1)] \quad (2.15) \\
y_m(k+3) &= g[y(k+2), y(k+1), y(k), u(k+2), u(k+1), u(k)] \\
y_m(k+4) &= g[y(k+3), y(k+2), y(k+1), u(k+2, u(k+2), u(k+1)]
\end{aligned}
$$

In order to compute $y_m(k+2)$, $y_m(k+3)$, $y_m(k+3)$ future process outputs and future values of the control signal (i.e. the decision variables of the MPC algorithm) are required. For NNARMAX model, future errors $[y(k+p) - y_m(k+p)]$ are additionally required. Since future process measurements are not available, the NNARX or NNARMAX can be used only for once step ahead prediction where the neural model output is a function of the current and past input-output process data.

The NNOE model (2.12) seems more adequate for multi-step ahead predictions because it is a completely recurrent structure depending only on the process inputs and the fed back model predictions:

$$
\begin{aligned}
y_m(k) &= g[y_m(k-1), y_m(k-2), y_m(k-3), u(k-1), u(k-2), u(k-3)] \\
y_m(k+1) &= g[y_m(k), y_m(k-1), y_m(k-2), u(k), u(k-1), u(k-2)] \\
y_m(k+2) &= g[y_m(k+1), y_m(k), y_m(k-1), u(k+1), u(k), u(k-1)] \\
y_m(k+3) &= g[y_m(k+2), y_m(k+1), y_m(k), u(k+2), u(k+1), u(k)] \\
y_m(k+4) &= g[y_m(k+3), y_m(k+2), y_m(k+1), u(k+2, u(k+2), u(k+1)]
\end{aligned}
$$

$$(2.16)$$

According to (2.16) the output of the NNOE model does not use process output measurements. However, in case of model inaccuracies and under parameterization, prediction errors can be accumulated and jeopardize the control system.

We propose here a predictive neural model that is a mixture of NNARX and NNOE and substitute the past model outputs with past process measurements. In the example above $y_m(k-1)$, $y_m(k-2)$, $y_m(k-3)$ in (2.16) will be substituted by $y(k-1)$, $y(k-2)$, $y(k-3)$. Then, the predictive form of (2.13) is

$$
y_m(k+p/k) = a_0 + \sum_{j=1}^{H} a_j \, \varphi\big(z_j(k+p/k)\big) \quad (2.17)
$$

The proposed modification (Model 1) is comparatively studied with the NNOE structure (Model 2).

2.3.4.1 On-line Multistep Ahead Predictive Model1

$$
\begin{aligned}
z_j(k + p/k) = b_{j0} &+ \sum_{i=1}^{\text{var}1} b1_{ji} y_m(k - i + p/k) \rightarrow \textbf{future model predictions} \\
&+ \sum_{i=1}^{n_y - \text{var}1} b2_{ji} y(k - i) \rightarrow (\textbf{measured(past)processoutputs}) \\
&+ \sum_{i=1}^{\text{var}2} c1_{ji} u(k - \tau + 1 - i + p/k) \rightarrow (\text{future input signals}) \\
&+ \sum_{i=1}^{n_u - \text{var}2} c2_{ji} u(k - \tau + 1 - i) \rightarrow (\text{past input signals})
\end{aligned}
$$

$$(2.18)$$

2.3.4.2 On-line Multistep Ahead Predictive Model 2 (NNOE)

$$
\begin{aligned}
z_j(k + p/k) = b_{j0} &+ \sum_{i=1}^{\text{var}1} b1_{ji} y_m(k - i + p/k) \rightarrow \textbf{future model predictions} \\
&+ \sum_{i=1}^{n_y - \text{var}1} b2_{ji} y(k - i) \rightarrow (\textbf{past model outputs}) \\
&+ \sum_{i=1}^{\text{var}2} c1_{ji} u(k - \tau + 1 - i + p/k) \rightarrow (\text{future input signals}) \\
&+ \sum_{i=1}^{n_u - \text{var}2} c2_{ji} u(k - \tau + 1 - i) \rightarrow (\text{past input signals})
\end{aligned}
$$

$$(2.19)$$

2.3.4.3 Neural Model Structure for Training and MPC Implementation

The training stage of any of the regression models (A–D) is usually based on the NN structure (2.13)–(2.14) where the objective is to minimize the Sum of the Squared Errors (SSE)

$$SSE - \sum_{k \in data\ set} [d(k) - y_m(k/k - 1)]^2 \qquad (2.20)$$

$y_m(k/k - 1)$ denotes the output of the neural model for the current sampling instant k calculated using signals up to the sampling instant $k-1$ and $d(k)$ is the process output collected during the identification experiment. Such training ignores the multi step ahead predictive role of the model (2.17) in the MPC framework. Due to the different neural model structures used for training and MPC implementation, prediction errors are further propagated.

In order to overcome this problem, during the training of Models 1 and 2, we use the same neural structure as the one used for prediction, Eqs. (2.18–2.19). The network input is provided with the process inputs and outputs from the identification experiments and the model predictions. A normalized form of (2.20) over the prediction horizon is used as a cost function. It is denotes as the Relative Mean Square (RMS) error

$$RMS = \sqrt{\frac{\sum_{p=1}^{H_p} (d(k+p) - y_m(k+p/k))^2}{\sum_{p}^{H_p} (d(k+p))^2}},$$

where $y_m(k + p/k)$ denotes the predictions of the model output for the future sampling instant $k + p$ calculated at the current sampling instant k and $d(k + p)$ is the process output collected during the identification experiment.

2.4 Identification Experiments

Data for NN training was collected according to the following experiments.

2.4.1 Classical (one test) Identification Experiment

Inputs (u_i) are generated as random signals. They are introduced to a dynamic crystallizer simulator used for laboratory experiments [24] and the respective process responses are recorded (y_i). The neural Models 1 and 2 are trained with data set $\{ u_i - u_{i,\text{mean}}, y_i - y_{i,\text{mean}} \}$, where ($u_{i,\text{mean}}, y_{i,\text{mean}}$) are the respective mean values of the collected input and output time series. This classical identification experiment is illustrated in Fig. 2.3 ($i = 1, 2$).

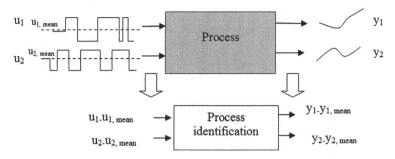

Fig. 2.3 Classical (one test) identification experiment

2.4.2 Double Test Identification

An alternative of the classical identification experiment was also studied. Prior to the test with randomly generated inputs (u_i), a test with constant inputs $\left(u_i' = u_{i,\,\mathrm{mean}}\right)$ is performed and the process reactions (y_i, y_i') are recorded (Fig. 2.4). Then the neural models are trained with data set $\left\{ u_i - u_i',\, y_i - y_i' \right\}$.

We expect that the second experiment has the potential to extract better the process dynamics.

Remark In case reference input trajectories are available (for example as a result of offline process design and optimization), the first test may be performed with these trajectories instead of constant inputs.

2.5 Error Tolerant MPC

2.5.1 Problem Formulation

In order to further relax the computational burden, a modification of the general NMPC problem (2.2), is proposed here with the following cost function

$$
\left. \begin{aligned}
\min_{\Delta u(k/k),\Delta u(k+1/k),\ldots\Delta u(k+H_c-1/k)} J = \\
\lambda_1 \sum_{p=1}^{H_p} \left(e(k+p/k)\right)^2 + \lambda_2 \sum_{i=1}^{H_c} \left(\Delta u(k+p/k)\right)^2
\end{aligned} \right\}, \quad if\ E_{\Sigma} > \alpha,\ \alpha \in R^+ \quad (2.21)
$$

$$
\text{where } E_{\Sigma} = \frac{1}{H_p} \sum_{p=1}^{H_p} |e(k+p/k)| \tag{2.22}
$$

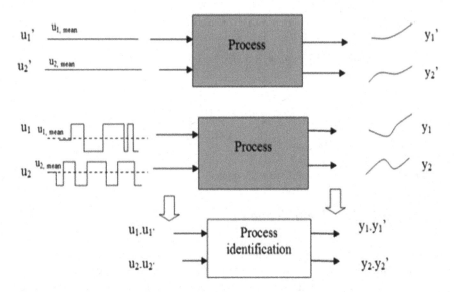

Fig. 2.4 Double-test identification experiment

subject to the same constraints (2.3–2.5). With the cost function (2.21) the future control actions are calculated as

$$u(k/k) = \begin{cases} u : \begin{cases} \displaystyle \min_{[\Delta u(k/k),\Delta u(k+1/k),.....\Delta u(k+H_c-1/k))]} J = \\ \displaystyle \lambda_1 \sum_{p=1}^{H_p} (e(k+p/k))^2 + \lambda_2 \sum_{p=1}^{H_c} (\Delta u(k+p/k))^2, E_{\sum} > \alpha \\ u^* \qquad\qquad\qquad\qquad\qquad\qquad if\ E_{\sum} \le \alpha \end{cases} \end{cases}$$

(2.23)

Equation (2.23) is a particular form of the general performance index defined by (2.2). We denote it as an error tolerant (ET) MPC because the optimization is performed only when the error function ($E_{\sum}$) is above a predefined real value α. When $E_{\sum}$ is inside the α-strip the control action is equal to u^*, the last implemented optimal control. The price to be paid is that the tracking is not achieved asymptotically, but in a neighborhood of the reference. However, the α tolerance can be arbitrarily small and is determined on a case by case basis, which suffices for practical purposes. The error function in (2.22) is defined as the mean of the errors between the predicted outputs and their reference values along the next H_p steps.

2.5.2 Selection of MPC Parameters

α is a design parameter whose choice is decisive for achieving a reasonable compromise between computational costs and tracking error. While a formal procedure for selecting is not defined, the error tolerance is chosen based on common sense consideration of $\pm 1 - \pm 5$ % error around the set-point.

The choice of H_p is related with the sampling period (Δt) of the digital control implementation, which in its turn is a function of the settling time t_s (the time before entering into the 5 % around the set-point) of the closed loop system. As a rule of thumb, it is suggested Δt to be chosen at least 10 times smaller than t_s, [18]. Hence, the prediction horizon can be chosen as $H_p = \mathbf{Int}\left(\frac{t_s}{\Delta t}\right)$. It is known that the smaller the sampling time, the better is the reference trajectory tracking and disturbance rejection. However, choosing a small sampling time yields a large prediction horizon. In order to compute the optimal control input, the optimization (2.2) is performed at each sampling time and requires large amount of computer memory per iteration and fast communication and computation resources. Such requirements are still unbearable for many industries with not-state-of-the-art control equipment. The ETMPC introduced by (2.23) has the potential to reduce the computational burden in such cases complementing other solutions in the literature for feasible real time optimal control.

In the MPC control tests (Sect. 2.8) the design parameter λ_1 is set to 1, while the choice of λ_2 is based on the following empirical expression

$$\lambda_2 = \frac{e_{\max} P / 100}{\left(u_{\max} - u_{\min}\right)^2} \tag{2.24}$$

where P defines the desired contribution of the second term in (2.21) (0 % $\leq$ P $\leq$ 100 %) and

$$e_{\max} = \max\left(\left(ref - y_{\max}\right)^2, \left(ref - y_{\min}\right)^2\right) \tag{2.25}$$

The intuition behind (2.24–2.25) is to compensate the magnitude orders of the two terms of (2.21).

2.5.3 Normalized ETMPC

A normalized version of the cost function (2.21) was implemented in the numerical tests

$$\min_{\Delta u(k/k),\Delta u(k+1/k),.....\Delta u(k+H_c-1/k)} J = \lambda_1 RMS + \lambda_2 ACE$$

$$\text{if } E_\Sigma = \frac{1}{H_p} RMS > \alpha, \ \alpha \in R^+ \tag{2.26}$$

where the Relative Mean Square (RMS) error is

$$RMS = \sqrt{\frac{\sum\limits_{p=1}^{H_p} (ref(k+p/k) - \hat{y}(k+p/k))^2}{\sum\limits_{i=1}^{H_p} (ref(k+p/k))^2}} \tag{2.27}$$

and the Average Control Effort (ACE)

$$ACE = \frac{\sum\limits_{p=1}^{H_c} (\Delta u(k+p/k))^2}{H_c} \tag{2.28}$$

The normalization prevents from abrupt changes of the cost function and makes a balance between the two terms. The intuition behind is that RMS incorporates the constraints that the controlled output should not deviate too far from the reference values and ACE represent the assumption that the input variations are sufficiently slow. The cost function in (2.26) is smoother and guarantees empirically the Bounded Input Bounded Output (BIBO) stability of the closed loop system.

2.6 Sugar Crystallization Case Study

2.6.1 Process Description

Sugar production is characterized by strongly non-linear and non-stationary dynamics and goes naturally through a sequence of relatively independent stages: charging, concentration, seeding, setting the grain, crystallization, tightening and discharge [24].

The feedback control policy is based on measurements of the flowrate, the temperature, the pressure, the stirrer power and the supersaturation (by a refractometer). Measurements of these variables are usually available for a conventional crystallizer.

Charging. During the first stage the crystallizer is fed with liquor until it covers approximately 40 % of the vessel height. The process starts with vacuum pressure of around 1 bar (equal to the atmospheric pressure) and reduces it up to 0.23 bar.

When the vacuum pressure reaches 0.5 bar, the feed valve is completely open such that the feed flowrate is kept at its maximum value. When the liquor covers 40 % of the vessel height, the feed valve is closed and the vacuum pressure needs some time to stabilize around the value of 0.23 bar before the concentration stage starts.

Concentration. The next phase is the concentration. The liquor is concentrated by evaporation, under vacuum, until supersaturation reaches a predefined value (typically 1.11). At this stage seed crystals are introduced into the pan to induce the production of crystals. This is the beginning of the third (crystallisation) phase.

Crystallisation (main phase). In this phase as evaporation takes place further liquor or water is added to the pan in order to guarantee crystal growth at a controlled supersaturation level and to increase total contents of sugar in the pan. Near to the end of this phase and for economical reasons, the liquor is replaced by other juice of lower purity (termed syrup). The supersaturation is the main driving force of this stage but the actual measured variable is the brix of the solution. However, due to a straightforward relation between supersaturation and brix, the supersaturation (S) is considered as the controlled process output.

Tightening. Once the pan is full the feeding is closed. The tightening stage consists principally in waiting until the suspension reaches the reference consistency, which corresponds to a volume fraction of crystals equal to 0.5. The stage is over when the stirrer power reaches the maximum value of 50 A. The steam valve is closed, the water pump of the barometric condenser and the stirrer are turned off. Now the suspension is ready to be unloaded and centrifuged.

The different phases are comparatively independent process states. The crystallization is the main stage responsible for the product quality quantified by the average (in mass) crystal particle size (AM) at the end of the process and the final coefficient of particle variation (CV). The present study is focused on defining an efficient MPC only for the crystallization phase. Based on a set of industrial data collected over normal white sugar production cycles, average values for the main process variables were determined and summarised in Table 2.1. Table 2.2 consists of the reference values and restrictions of the process quality variables evaluated at the batch end. For more details with respect to the process see [25].

2.6.2 Crystallization Macro Model

The general phenomenological model of the fed-batch crystallization process consists of mass, energy and population balances. While the mass and energy balances are common expressions in many chemical process models, the population balance is related with the crystallization phenomenon which is still an open modelling problem.

Table 2.1 Process variables

Name	Notation	Average value	Max value
Liquor/Syrup feed flowrate	F_f	0.0057 m³/s	0.025 m³/s
Steam flowrate	F_s	1.6 m³/s	2.75 kg/s
Feed temperature	T_f	65 °C	–
Steam temperature	T_s	140 °C	–
Brix of feed	Bx_f	0.7	–
Steam pressure	P_s	2 bar	–
Temperature of massecuite	T_m	72.5 °C	–
Vacuum pressure	P_{vac}	0.25 bar	0.5 bar
Brix of the solution	Bx_{sol}	2 bar	–
Stirring power	W	25 A	50 A

Table 2.2 Hard constraints of process quality variables

Name	Notation t_{final}–final time	Value
Average (in mass) crystal size	$AM(t_{final})$	0.55–0.6 mm (ref.)
Coef. of variation	$CV(t_{final})$	below 32 %
Volume	$V(t_{final})$	35 m³ (max)
	S	
Supersaturation		1.3 (max)

2.6.2.1 Mass Balance:

The mass of all participating solid and dissolved substances are included in a set of conservation mass balance equations

$$\dot{M} = f(M(t), F(t), P1), \quad 0 \le t \le t_f, \quad M(0) = M_0 \qquad (2.29)$$

where $M(t) \in R^q$ and $F(t) \in R^m$ are the mass and the flow rate vectors, with q and m dimensions respectively, and t_f is the final batch time. $P1$ is the vector of physical parameters as density, viscosity and purity.

2.6.2.2 Energy Balance:

The general energy (E) balance model is

$$\dot{E} = f(E(t), M(t), F(t), P2), \quad 0 \le t \le t_f, \quad E(0) = E_0 \qquad (2.30)$$

where $P2$ incorporates the enthalpy terms and specific heat capacities derived as functions of physical and thermodynamic properties.

2.6.2.3 Population Balance:

Mathematical representation of the crystallization rate can be achieved through basic mass transfer considerations or by writing a population balance represented by its moment equations [26]. Employing a population balance is generally preferred since it allows to take into account initial experimental distributions and, most significantly, to consider complex mechanisms such as those of size dispersion and/or particle agglomeration/aggregation. Hence

$$\dot{\eta}_i = f\left(\eta_i(t), \tilde{B}_0, G, \beta'\right), \quad 0 \leq t \leq t_f, \quad i = 0, 1, 2, \ldots \eta_i(0) = \eta_{i0} \qquad (2.31)$$

where η_i is the j-th moment of the mass-size particle distribution function, $\tilde{B}_0$, G and β' are the kinetic variables nucleation rate, linear growth rate and the agglomeration kernel, respectively. The process quality measures are derived as

$$AM(t) = \eta_1(t)/\eta_0(t) \qquad (2.32)$$

$$CV(t) = \left(\eta_0(t)\eta_2(t)/\eta_1^2(t) - 1\right)^{1/2} \qquad (2.33)$$

The control objective is to get a desired final average crystal size $AM(t_{final})$ = AM_{ref} = 0.55 and to minimize $CV(t_{final})$.

2.7 Neural Model Identification

2.7.1 Identification Experiments

Based on the macro model (2.29–2.33), a dynamical process simulator was developed [24]. It consists of mass, energy and population balances and was extensively tuned with real data from two industrial units (Sugar refinery RAR.SA, Portugal and Company 30 de Noviembre, Pinar del Río, Cuba). For all measured process variables (the vacuum pressure, brix and temperature of the feed flow, pressure and temperature of the steam), the simulator provides an option to introduce industrial measurements, which serves as a natural source of disturbance and noise. The results of the MPC tests are related with the Sugar refinery RAR.SA plant.

Over the crystallization stage the controlled variable is the supersaturation (S). The manipulated variables are the feed flow rate (F_f), during the first part of the crystallization and latter on the steam flow rate (F_s). The open-loop input-output process data were acquired either by available industrial measurements (batch 1, 2, 3) or by generated inputs as random signals limited by maximum of 10 % around the average values collected in Table 2.1. In both cases the sampling period is $t_s = 10$ s.

Data bases for NN training and testing were obtained according to the single or double test experiments discussed in Sect. 2.4.

2.7.2 NN Structure Selection

In term of NN structure selection, the number of delayed signals (n_y and n_u) used as regressors (lag space) and the number and type of hidden units need to be determined.

2.7.2.1 Lag Space Selection

Insufficient lag space means that essential dynamics is not modeled. Too many regressors imply overestimated model order. A procedure based on the Lipschitz quotient is used to optimize the lag space, assuming that the process can be represented accurately by a function that is reasonable smooth in the regressors. A detailed description of this approach can be found in [13] and [27]. The procedure is the following:

1. For a given choice of the lag space n_y determine the Lipschitz quotients for all combinations of input-output pairs

$$q_{ij} = \left| \frac{y_i^{NN}(k) - y_j^{NN}(k)}{x_i(k) - x_j(k)} \right|, \quad i \neq j, \quad i,j = 1, 2, \ldots . N \qquad (2.34)$$

where ‖ specifies the Euclidian norm, i.e. the distance, N is the number of available samples, $x_i(k)$ is one element of the input vector $x(k)$

$$y^{NN}(k) = g[x(k), \theta], \quad x(k) = [x_1, x_2, \ldots . x_z] \qquad (2.35)$$

The Lipschitz condition states that q_{ij} is always bounded if the function g is continuous.

2. Select the p largest quotients, where $p = 0.01 \, N \div 0.02 \, N$
3. Evaluate the criterion

$$q = \left(\prod_{i=1}^{p} \sqrt{n_y} q_i^{(n_y)} \right)^{\frac{1}{p}} \qquad (2.36)$$

4. Repeat (1–3) for a number of different lag structures.
5. Plot the criterion q as a function of the lag space and select the optimal number of regressors as the "knee point" of the curve.

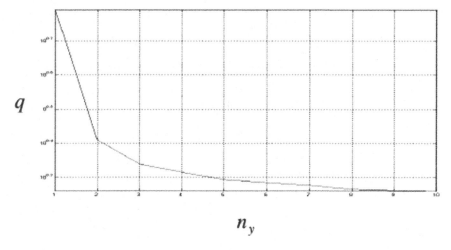

Fig. 2.5 Criterion (2.36) as a function of the lag space n_y

For the crystallization case

$$x(k) = \lfloor y(k-1), \ldots y(k-n_y), u(k-\tau), \ldots u(k-\tau-n_u+1) \rfloor \qquad (2.37)$$

The process is simplified by choosing $n_y = n_u$. The crystallization stage takes about 45–50 min., with a sampling rate $t_s = 10$ s., hence $N \approx 300$ samples and $p = 3 \div 6$. Based on the results depicted in Fig. 2.5, third order dynamical model was defined.

2.7.2.2 Number of Hidden Nodes

The studied neural models have equal input arguments $n_y = n_u = 3$, $\tau = 1$, and hyperbolic tangent neurons in the hidden layer. Table 2.3 summarized the results with respect to neural model training and testing as a function of the number of hidden nodes (H). Classical training algorithms have been tested: steepest descent (SD), conjugate gradient (CG) [28] and Levenberg-Marquardt (LM) method [29]. In terms of RMS, the methods are similar, however the LM method converges most rapidly. Hence, the LM training suits better for on-line neural model tuning or periodical on-line model calibration.

More hidden nodes H (3, 4, 5, 6, 7), lower RMS for the training data, however for H > 6, the RMS for test data rapidly increases (overfitting problem). Therefore H = 6 is chosen as a compromise between accuracy and complexity.

Table 2.3 The effect of the number of hidden nodes and the training algorithm on the accuracy of the neural dynamical model

H	No. of weights	Training method	RMS training	RMS test
3	19	SD	0.195	0.289
		CG	0.189	0.296
		LM	0.201	0.311
4	25	SD	0.087	0.134
		CG	0.141	0.199
		LM	0.109	0.220
5	31	SD	0.079	0.097
		CG	0.056	0.261
		LM	0.191	0.301
6	37	SD	0.068	0.091
		CG	0.011	0.101
		LM	0.026	0.099
7	43	SD	0.066	0.172
		CG	0.009	0.391
		LM	0.017	0.283

2.7.2.3 NN Pruning

Besides the number of regressors and the number of hidden nodes, the number of connections among the neurons is also an important NN parameter. The Optimal Brain Damage (OBD) pruning algorithm [30] is used to further improve the ratio between accuracy and complexity

OBD determines the optimal network architecture by removing the superfluous weights from the network in order to avoid the overfitting of the data by the NN. The objective is to find a set of weights whose removing is likely to result in the least change in RMS. The saliency S_i of weight ϑ_i is defined as a function of the second-order derivative of the cost function with respect to $\vartheta_i \left(\frac{\partial^2 RMS}{\partial(\theta_i)^2} \right)$

$$S_i = \frac{\partial^2 RMS}{\partial(\theta_i)^2}(\theta_i)^2, \quad i = \{1, 2, \ldots n\} \quad n = (n_y + n_u + 1)H + H + 1 \quad (2.38)$$

where n is the total number of network weights and

$$\vartheta = [\theta_1, \theta_2, \ldots \theta_n] = [\{a_j\}, \{b_{ji}\}, \{c_{jk}\}],$$
$$\text{for} \quad j = 0, \ldots H, \ i = 0, \ldots n_y, k = 1, \ldots n_u$$

Table 2.4 The effect of NN prunning on the final process quality measures

Method	No weights	AM_final (mm) (ref. 0.56)	CV_final (%)
NN MPC	37	0.615	30.28
NN ETMPC	(before	0.603	31.14
Normalized NN ETMPC	pruning)	0.573	29.36
NN MPC	23	0.637	30.26
NN ETMPC	(after	0.636	30.42
Normalized NN ETMPC	pruning)	0.550	28.74

Table 2.5 MPC design parameters

t_s (s) Settling time	Δt (s) Sampling period	H_p Prediction horizon	H_c Control horizon	λ_2 Weight	Set-point	α-strip (1 %)
Control variable-feed flowrate F_f; controlled variable-supersaturation S						
40	4	10	4	0.1	1.15	0.01
PID: proportional gain $k_p = -0.5$; integral time const $\tau_i = 40$; $\tau_d = 0$						
Control variable-feed flowrate F_s; controlled variable-supersaturation S						
60	4	15	4	0.01	1.15	0.01
PID: proportional gain $k_p = 20$; integral time const $\tau_i = 40$; $\tau_d = 0$						

The OBD pruning algorithm can be summarized as follows:

1. Train the fully connected NN
2. Compute the saliency (2.38) with respect to all weights
3. Eliminate the weight with the smallest saliency value
4. Retrain the NN
5. Stop if pruning leads to higher RMS
6. If not, go to step 2.

The NN started with 6 hidden nodes and 37 weights. After the OBD pruning procedure, 14 weights were removed but the network predictions of the end-point process quality measures AM_final and CV_final are still reliable (see Table 2.4).

2.8 NN ETMPC Control Tests

The NN-based MPC was tested on a dynamic crystallizer simulator [24]. The MPC design parameters are summarized in Table 2.5. The NN MPC with the classical cost function (2.2) and the Error Tolerant modifications (not normalized and normalized NN ETMPC) with cost functions (2.21) and (2.26) respectively, are compared with a PID controller designed in the following velocity form

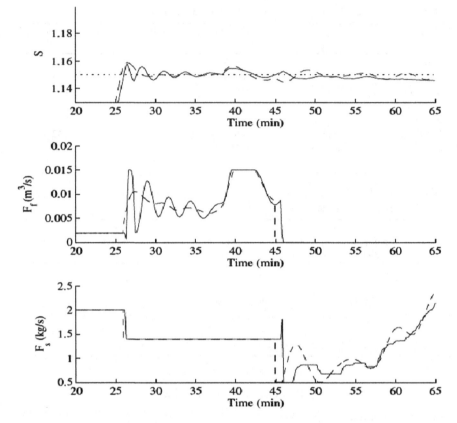

Fig. 2.6 Controlled variable (S-supersaturation), manipulated variables (Ff-feed flowrate), Fs-steam flow rate. *Dashed line—NN-Normalized ETMPC; Full line—NN-MPC*

$$u(k) = u(k-1) + K_p\left((e(k) - e(k-1)) + \frac{\Delta t}{\tau_i} \cdot e(k) \right.$$
$$\left. + \frac{\tau_d}{\Delta t} \cdot (e(k) - 2 \cdot e(k-1) + e(k-2))\right) \tag{2.39}$$

The optimized PID parameters k_p, τ_i, τ_d are summarized in Table 2.5. The tuning suggests that the derivative time constant is not necessary therefore $\tau_d = 0$.

2.8.1 Set-Point Tracking

Time trajectories of the controlled and the manipulated variables only for the crystallization stage are depicted in Figs. 2.6, 2.7, 2.8. During the first half of the stage liquor is introduced into the pan and its flow rate (F_f) is the way to control

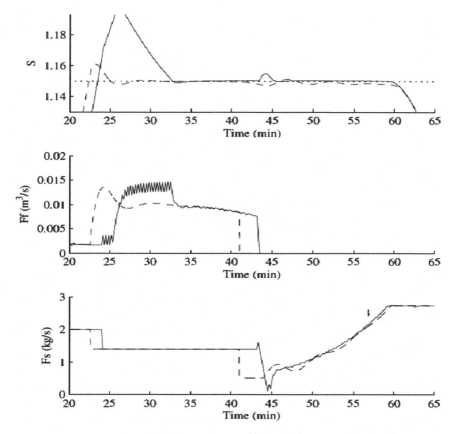

Fig. 2.7 Controlled variable (S-supersaturation), manipulated variables (Ff-feed flowrate), Fs-steam flow rate. *Dashed line*—NN-Normalized ETMPC; *Full line*—NN-ETMPC

the supersaturation around the set point. When the complete amount of liqueur is added, the process is next controlled by the steam flow rate (F_s).

All MPC control loops show similar tracking performance. There results demonstrate that the explicitly introduced error tolerance does not worsen the output reference tracking. The price to be paid is that the (not normalized) NN ETMPC needs longer time to enter into the α-strip, however, it is not the case with the normalized NN ETMPC. The PID exhibits the worse tracking performance (Fig. 2.8—full line) which is due to the inherent nonstationarity of the process, while the PID parameters are tuned assuming stationary process parameters.

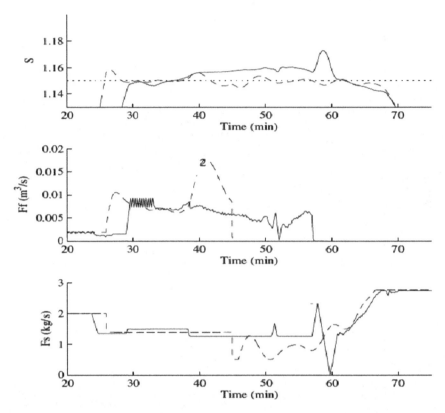

Fig. 2.8 Controlled variable (S-supersaturation), manipulated variables (Ff-feed flowrate), (Fs-steam flow rate). *Dashed line*—first principles model-Normalized ETMPC; *Full line*—PID

2.8.2 Computational Time Reduction

The principal contribution of the NN-based Error Tolerant MPC is with respect to the computational time reduction. Figures 2.9 and 2.10 depict the optimization time per iteration over one production cycle, comparing the classical MPC (Fig. 2.9a) with tree ETMPC alternatives: i) ETMPC with neural Model 1, Eq. (2.18); ii) ETMPC with neural Model 2, Eq. (2.19); iii) Normalized ETMPC with neural Model 1. The NN-based ETMPCs require significantly less time to compute the feasible (sub) optimal control actions. The average CPU times per iteration are summarized in Table 2.6. Note that they range from about 0.211–0.192 s for the classical MPC to 0.061–0.078 s for the Normalized NN-ETMPC, assuming only 1 % error tolerance. In case the process tolerates higher tracking imprecision the computational time can be further reduced.

Fig. 2.9 Optimization time per iteration. **a** NN-MPC **b** NN (Model 1)-ETMPC

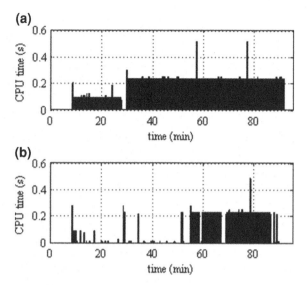

Fig. 2.10 Optimization time per iteration. **a** NN (Model 2)-ETMPC; **b** Normalized NN (Model 1)-ETMPC

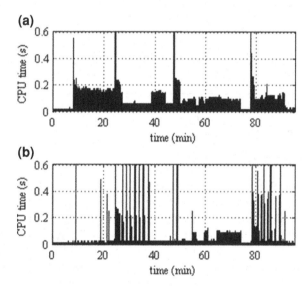

2.8.3 Final Product Quality

For practical applications it is interesting to study the influence of the NN predictive models on the process final performance. Therefore, their prediction quality is tested in the framework of the normalized NN ETMPC towards the principal control objectives, namely how close is the average crystal size (AM_final) to the desired value of 0.55 mm. and how variable is this size (CV_final) at the end of the process. Tables 2.7 and 2.8 collect the performed tests. First (Table 2.7), Models 1

Table 2.6 Average computational cost

Control method	NN Model	Average optim. time per iteration (S)
Normalized NN ETMPC	Model 1/eq. (2.18)	0.061
	Model 2/eq. (2.19)	0.078
NN ETMPC	Model 1	0.093
	Model 2	0.091
NN MPC	Model 1	0.192
	Model 2	0.211

Table 2.7 NN prediction quality in the framework of the normalized ETMPC

Model	Training data base	AM_final (mm) (ref. 0.55)	CV_final (%)
Model 1	1 batch (273 p.)	0.615	31.14
	2 batches (551 p.)	0.592	30.93
	3 batches (816 p.)	0.550	28.74
Model 2	1 batch (373 p.)	0.61	29.18
	2 batches (551 p.)	0.575	32.06
	3 batches (816 p.)	0.601	30.10

Table 2.8 Process model identification assuming model 1 structure

Identif.	Training data base	RMS test	AM_fin (mm) (ref. 0.55)	CV_final (%)
One test	Generated signals	0.256	0.644	30.19
	Industrial measurements	0.269	0.587	32.01
Double test	Generated signals	0.097	0.591	29.06
	Industrial measurements	0.102	0.550	28.74

and 2 were trained with industrial data acquired over one, two or three in spec batches. Increasing the training data influences the prediction quality mainly of Model 1 whose structure depends directly on the process measurements. In case less training data are available Model 2 exhibits better prediction ability. The effect of the identification strategies discussed in Sect. 2.4 is studied assuming an MPC controlled process with Model 1 predictions (Table 2.8). The double test identification is less sensitive with respect to the nature of the training data (generated signals or industrial measurements) and ends up with lower RMS.

2.9 Conclusions

The traditional position of many practitioners is still in favor of linear control solutions. However, for significantly nonlinear processes, as for example in the crystallization industry, the advantages of data-based process modeling, optimization and control are difficult to neglect. This chapter discusses the application of NNs as predictive models in the framework of nonlinear model based predictive control (MPC).

The proposed recurrent multistep ahead predictive neural model fits well to the MPC objectives and guarantees the same performance as the traditional models obtained from physical laws. Furthermore, the multi step ahead prediction role of the model in the MPC framework is taken into account during the NN training. These modifications, of the classical regression predictive models, reduce the propagation of the prediction error and have the advantage to be a straightforward modeling technology.

The second contribution of the present work is the heuristic modification of the MPC, termed ETMPC, where the optimization is executed only when the tracking error is above a predefined level. In applications where the tracking is allowed to tolerate some margins, the suboptimal NN-based ETMPC control can relax substantially the computational burden. In the presence of limited computational resources suboptimal control is often the compromise due to feasibility problems or lack of solution (convergence) within the restricted time per iteration.

From a systems engineering point of view, the crystallization phenomena can be found in a significant number of industrial processes (e.g. pharmaceutical industry, precipitation, wastewater treatment). Therefore, the control solution proposed for the present industrial case has the potential to be easily extended to other processes.

Acknowledgements The work was partially funded by the Portuguese National Foundation for Science and Technology (FCT) in the context of the project FCOMP-01-0124-FEDER-022682 (FCT reference PEst-C/EEI/UI0127/2011) and the Institute of Electrical Engineering and Telematics of Aveiro (IEETA), Portugal.

References

1. Clarke, D.W., Mohtadi, C., Tuffs, P.S.: Generalized predictive control. part 1. The basic algorithm. part 2: Extensions and interpretations. Automatica **23**, 137–148 (1987)
2. Rossiter J.A.: Model-based predictive control: a practical approach. CRC press, Boca Raton. (2003) ISBN 0-9493-1291-4
3. Morari, M.: Advances in model-based predictive control. Oxford University Press, Oxford (1994)
4. Balasubramhanya, L.S., Doyle III, F.J.: Nonlinear model-based control of a batch reactive distillation column, J. Process Control. **10**, 209–218 (2000)
5. Nagy, Z.K., Braatz, R.D.: Robust nonlinear model predictive control of batch processes. AIChE J. **49**, 1776–1786 (2003)
6. Nagy, Z.K.: Model based control of a yeast fermentation bioreactors using optimally designed artificial neural networks. Chem. Eng. J **127**(1), 95–109 (2007)
7. Qin, S.J., Badgewell, T.: A survey of industrial model predictive control technology. Control Eng. Pract **11**, 733–764 (2003)
8. Hunt, K.J., Sbarbaro, D., Zbikowski, R., Gawthrop, P.J.: Neural networks for control systems-a survey. Automatica **28**, 1083–1112 (1992)

9. Nørgaard, M.: System identification and control with neural networks. PhD thesis, Department of Automation, Technical University of Denmark, Lyngby (1996)
10. ŁAwryńCzuk, M.: A family of model predictive control algorithms with artificial neural networks. Int J. Appl Math Comput. Sci. **17**(2), 217–232 (2007)
11. ŁAwryńczuk, M., Tatjewskin, P.: Nonlinear predictive control based on neural multi models, Int. J. Appl. Math. Comput. Sci. **20**(1), 7–21 (2010)
12. Narendra, K.S., Parthasarathy, K.: Identification and control of dynamical systems using neural networks. IEEE Trans. Neural Networks **1**(1), 4–27 (1990)
13. Nørgaard, M., Ravn, O., Poulsen, N.K., Hansen, L.K.: Neural networks for modelling and control of dynamic systems. Springer, London (2000)
14. Morari, M., Lee, J.H.: Model predictive control: Past, present and future. In: Proceedings of the PSE'97-ESCAPE-7 Symposium, Trondheim (1997)
15. Nagy, Z.K., Mahn, B., Franke, R., Allgöwer, F.: Nonlinear model predictive control of batch processes: An industrial case study, IFAC World Congress, Prague (2005)
16. Lightbody, G., Irwin, G.W.: Nonlinear control structures based on embedded neural system models. IEEE Trans. Neural Networks **8**(3), 553–567 (1997)
17. Hunt, K.J. Sbarbaro, D.: Neural networks for nonlinear internal model control".IEE Proceedings-D **138**(5), 431–438 (1991)
18. Georgieva, P., Feyo de Azevedo, S.: Novel computational methods for modeling and control in chemical and biochemical process systems. In: do Carmo Nicoletti, M., Jain, L.C. (eds.) Comp. Intelligence Techniques for Bioprocess Modelling, Supervison and Control, SCI 218, Springer, Heidelberg, 99–125 (2009)
19. Parisini, T., Sanguineti, M., Zoppoli, R.: Nonlinear stabilization by receding-horizon neural regulators. Int. J. Control **70**(3), 341–362 (1998)
20. Lu, C.H., Tsai, C.C.: Adaptive predictive control with recurrent neural network for industrial processes: An application to temperature control of a variable-frequency oil-cooling machine. IEEE Trans. Industr. Electron. **55**(3), 1366–1375 (2008)
21. Peng, H., Yang, Z.J., Gui, W., Wu, M., Shioya, H., Nakano, K.+ : Nonlinear system modeling and robust predictive control based on RBF-ARX model. Eng. Appl.Artif. Intell. **20**(1), 1–9 (2007)
22. Hagan, M.T., Demuth, H.B., Beale, M.H.: Neural Network Design. PWS Publishing Company, Pacific Grove (1996)
23. Mandic, D.P., Chambers, J.A.: Recurrent Neural Networks for Prediction: Learning Algorithms, Architectures and Stability (Adaptive & Learning Systems for Signal Processing, Communications & Control), Wiley, New York (2001)
24. Georgieva, P., Meireles, M.J., Feyo de Azevedo, S.: Knowledge-based hybrid modelling of batch crystallization when accounting for nucleation, growth and agglorameration phenomena. Chem. Eng. Sci. **58**, 3699–3713 (2003)
25. Oliveira, C., Georgieva, P., Rocha, F., Feyo de Azevedo, S.: Artificial neural networks for modeling in reaction process systems. Neural Comput. Appl. **18**, 15–24 (2009)
26. Paz Suárez, L.A., Georgieva, P., Feyo de Azevedo, S.: Nonlinear MPC for fed-batch multiple stages sugar crystallization, IChemE. Chem. Eng. Res. Des. (2010). doi:10.1016/j.cherd. 2010.10.010
27. He, X., Asada, H.: A new method for identifying orders of input-output models for nonlinear dynamic systems, In: Proceedings of the American Control Conference, San Diego (1993)
28. Zhou, G., Si, J.: Advanced neural-network training algorithm with reduced complexity based on jacobian deficiency. IEEE Trans. Neural Networks **9**(3), 448–453 (1998)
29. Lera, G., Pinzolas, M.: Neighborhood based levenberg-marquardt algorithm for neural network training. IEEE Trans. Neural Networks **13**(5), 1200–1203 (2002)
30. Reed, R.: Pruning algorithms—a survey. IEEE Trans. Neural Networks **4**(5), 740–747 (1993)

Chapter 3
Advances in Multiple Models Based Adaptive Switching Control: From Conventional to Intelligent Approaches

Nikolaos A. Sofianos and Yiannis S. Boutalis

Abstract The scope of this chapter is to trace the recent developments in the field of Intelligent Multiple Models based Adaptive Switching Control (IMMASC) and provide at the same time all the essential information about the conventional single model and multiple models adaptive control, which constitute the base for the development of the new intelligent methods. This work emphasizes on the importance and the advantages of IMMASC in the field of control systems technology presenting control structures that contain linear robust models, neural models and T-S (Takagi-Sugeno) fuzzy models. One of the main advantages of switching control systems against the single model control architectures is that they are able to provide stability and improved performance in multiple environments when the systems to be controlled have unknown parameters or highly uncertain parameters. Some hybrid multiple models control architectures are presented and a numerical example is given in order to illustrate the efficiency of the intelligent methods.

3.1 Introduction

Multiple models logic in control systems has been inspired from the necessity to describe and control as good as possible a system which is time-varying, uncertain or unknown. Instead of using one model with a lot of uncertainty in its dynamic equation, we use unique different models to describe the system in various operational conditions. This method reduces the conservativeness of adding

N. A. Sofianos · Y. S. Boutalis (✉)
Department of Electrical and Computer Engineering, Democritus University of Thrace, 67100 Xanthi, Kimmeria, Greece
e-mail: ybout@ee.duth.gr

N. A. Sofianos
e-mail: nsofian@ee.duth.gr

V. E. Balas et al. (eds.), *Innovations in Intelligent Machines-5*,
Studies in Computational Intelligence 561, DOI: 10.1007/978-3-662-43370-6_3,
© Springer-Verlag Berlin Heidelberg 2014

uncertainty into a single model and provides more realistic descriptions for the real systems. The term "multiple" introduced for the first time in the control systems theory in the 1970s when multiple Kalman filters were used to improve the accuracy of the state estimate in control problems [1, 2]. The substantial difference of these methods with the methods that developed in the 1980s was that no switching was involved in the control procedure and there were no stability guarantee. Martensson [3], introduced switching in the framework of adaptive control. Meanwhile, direct switching and indirect switching were proposed as two new techniques. In direct switching which is not very realistic and useful in real problems [4, 5] the output of the system defines the time of the switching to the next predefined controller. On the other hand indirect switching [6] uses multiple models to determine both when and to which controller to switch. It is obvious that indirect switching was far more reliable than direct switching because the sequence of the controllers was not pre-determined but it was determined according to the real conditions of the system. In [7] there was a modification in the previous approaches and the hysteresis switching algorithm was reexamined in a broader context. In [8] the author used multiple fixed models for switching and control and in [9, 10] the authors introduced the switching and tuning of the multiple models. These works were the base for the significant developments in the field of multiple models control during the last twenty years. In [11] the authors concentrate their efforts to switching and they proposed a scheme which is supervised by a high-level switching logic, providing very promising results. In [12, 13], performance improvement methods based on parameter switching was applied for back-stepping and sliding mode type controllers respectively. Transient performance improvement for a class of systems along with a quantitative evaluation was investigated in [14]. Ye [15] used only adaptive identification models in order to control nonlinear systems in parametric-strict-feedback form and provided a model convergence proof. Recently in [16, 17] the authors proposed the adaptive mixing control method that makes available the use of the full suite of powerful design tools from LTI theory and also they combined the controller mixing strategy with a multiple parameter estimation architecture plus a hysteresis switching logic in order to eliminate the dependence on the initial conditions of the parameter estimator. Finally in the field of conventional multiple models control there are some new works [18, 19] which focus on the number of adaptive estimation models and the source of information for every model leading in a faster convergence and in better performance than other techniques.

Another very promising part of multiple models control concerns the use of some intelligent approaches such as neural networks and fuzzy systems. These approaches are very useful especially when the system to be controlled is nonlinear or the mathematical expression of the system does not exist or it is very complicated. In Chen and Narendra [20] the authors propose an adaptive control scheme for a class of nonlinear discrete-time single-input-single-output (SISO) dynamical systems with boundedness of all signals by using a linear robust adaptive controller and a neural network based nonlinear adaptive controller. By using a switching scheme the control signal is determined at every time instant by the most suitable controller. The linear controller assures boundedness of all signals but not a satisfactory performance and is useful especially at the first seconds of the control procedure, when the adaptive neural network based controller is not well initialized and needs to be trained. In [21] an adaptive multiple neural network controller (AMNNC) with a supervisory controller for a class of uncertain nonlinear dynamic systems was developed. The AMNNC is a kind of adaptive feedback linearizing controller where nonlinearity terms are approximated with multiple neural networks. The weighted sum of the multiple neural networks is used to approximate system's nonlinearity for the given task. Each neural network represents the system dynamics for each task and with the help of a supervisory controller, the resulting closed-loop system is globally stable in the sense that all signals involved are uniformly bounded. Fu and Chai [22] propose a neural network and multiple model based nonlinear adaptive control approach for a class of uncertain nonlinear multivariable discrete-time dynamical systems. The controller architecture is similar to [20] but here the authors extended it to multiple-input-multiple-output (MIMO) systems and relaxed the assumption of global boundedness on the high order nonlinear terms.

On the other side of intelligent methods, fuzzy logic based control techniques entered the world of multiple models for the first time in [23–25]. In [23] the authors make use of some adaptive fuzzy identification models along with their adaptive fuzzy controllers. The switching between the controllers is based on a cost criterion and the adaptive laws are obtained by using the Lyapunov theory. The control philosophy is the same in [24], but there, the stability analysis contains a minimum time between the switchings and a hysteresis. Also a convergence analysis is given. In [25] there is a modification in the control scheme. Fixed, free adaptive and reinitialized adaptive models are used in order to enhance the transient response of the controlled system. These methods are very promising and new horizons are opened for fuzzy logic in the framework of multiple models.

The rest of the chapter is organized as follows. In Sect. 3.2, an overview of some conventional control methods using fixed and adaptive multiple models is given along with some basics on adaptive control. In Sect. 3.3, the use of neural models in multiple models control is presented. The role of T-S fuzzy models in multiple models switching control is given in Sect. 3.4. A numerical example is presented in Sect. 3.5 and finally the conclusions are given in Sect. 3.6.

3.2 Single Model Adaptive Control and Multiple Models Switching Adaptive Control

Adaptive control theory and multiple models switching control are two different methods that can be combined with great success. In the first subsection there is an overview from conventional to intelligent adaptive control approaches while in the second subsection the basic methods of conventional multiple models control are given.

3.2.1 Adaptive Control Basics

Since the early 1950s adaptive control techniques [26] have been used with great success in the control systems field. Researchers have been developed many different and reliable approaches based on the idea of adaptation (see in [27] and the references there in). The motivation for the development of that kind of control was the difficulties that came up when the plant to be controlled was a rapidly time-varying system (airplanes, industrial systems, etc.). The classic control techniques such as robust control, predictive control etc. although they manage to control efficiently the vast majority of the time-varying systems, they end up to be very conservative in many cases. On the other hand, adaptive control techniques try to identify online the plant to be controlled—estimate its parameters—and they adapt the controller's parameters appropriately. This procedure is repeated at every time instant, ensuring the validity and the adaptiveness of the control signal. There are two main approaches in adaptive control, depending on the way the parameter estimator is combined with the control law. In the first approach which is referred to as indirect adaptive control, the plant parameters are estimated on-line and used to calculate the controller's parameters. In the second approach, referred to as direct adaptive control, the plant model is parameterized in terms of the controller's parameters which are estimated directly without intermediate calculations involving plant parameters estimates. These two approaches constitute the base for all the adaptive control techniques since 1950s where Kalman [28] suggested an adaptive pole placement scheme based on the optimal linear quadratic problem and Osburn [29] suggested a model reference adaptive control (MRAC) scheme. The stability theory based on Lyapunov approach was introduced by Parks [30], and some stable MRAC schemes using the Lyapunov design approach were designed and analyzed in [31, 32], adding a value in the adaptive control effectiveness. During the 1980s the term "robust" dominated in the adaptive control techniques ensuring the stability of the systems under unmodeled dynamics and bounded disturbances [33]. Meanwhile, some intelligent control methods such as fuzzy control and neural control started to play a key role in the adaptive control schemes. Wang [34] presented an adaptive fuzzy controller for affine non-linear systems with unknown functions. In this work, the unknown nonlinear functions are represented by fuzzy basis function based fuzzy systems and the parameters of the fuzzy systems including membership functions are updated according to some adaptive laws which are based on Lyapunov stability theory. The very promising results motivated the researchers to develop new methods in the adaptive fuzzy control field [35–44]. The common feature in these works is that they use fuzzy systems to approximate the unknown nonlinear functions of the nonlinear plants and represent the fuzzy systems in a suitable form i.e. linear regression with respect to the unknown or uncertain parameters and then they use the classical adaptive control methods to achieve the desired result. In addition some researchers imposed robust control techniques in fuzzy adaptive control due to the fact that the mismatches between the real model and its fuzzy model may lead to instabilities if these mismatches are not be taken into account. Other researchers tried to improve the performance or reduce the number of tuning parameters in the adaptive fuzzy control schemes [45] and others used genetic algorithms in order to tune the fuzzy membership functions and enhance the control scheme [46]. In [47] the authors provide a comparison between conventional adaptive control and adaptive fuzzy control, concluding that under certain circumstances the last can give better results.

Similar to the intensive interest about adaptive fuzzy control the researchers developed very intensively some adaptive techniques based on neural networks. Neural networks were mostly used as approximation models for the unknown nonlinearities of the plant functions due to their inherent approximation capabilities. This is very useful especially in cases where the plant modeling is very complicated. Adaptive neural control schemes based on Lyapunov stability theory have been proposed since 1990s for both nonlinear systems with certain types of matching conditions [48–50] and nonlinear triangular systems without the requirement of matching conditions [51–53]. In these approaches the stability of the closed loop system is guaranteed and the performance and robustness properties are readily determined. During the last decade many new approaches have been developed concerning perturbed strict feedback nonlinear systems [54], MIMO nonlinear systems [55, 56], nonlinearly parametric time-delay systems [57, 58], affine nonlinear systems [59] and non-affine pure-feedback non-linear systems [60].

Fuzzy adaptive control and neural adaptive control used together in hybrid schemes providing very promising results. These two control techniques complement to each other, that is neuro control provides learning capabilities and high computation efficiency in parallel implementation while fuzzy control provides a powerful framework for expert knowledge representation. The advantages of both fuzzy and neural control systems combined in neuro-fuzzy adaptive control techniques [61–66]. It is obvious from the aforementioned works that adaptive control and especially intelligent adaptive control is suitable for controlling highly uncertain, nonlinear, and complex systems. In the following subsection an overview of the multiple models control methods is given and the crucial role of adaptive multiple models control becomes more clear.

3.2.2 An Overview of Multiple Models Switching Control Methods

In this subsection the basic advantages of multiple models switching control over the single model adaptive control are mentioned and an overview of the most important conventional methods of this kind of control is given.

3.2.2.1 Why Multiple Models? From Adaptive to Multiple Models Adaptive Switching Control

The necessity for another kind of control rather than the classical adaptive control techniques can be found in the instability problems that arise when the system to be controlled is time variant and its parametric uncertainty is very high [67]. Classical adaptive control techniques have been designed to provide very good results when the plant is time invariant and the parametric uncertainty is small. But these ideal conditions are not realistic some times, especially in fields like biology, economics, engineering and social systems. In these fields the objective is to formulate an adaptive controller that could be tuned appropriately in order to take the right decision in the right time while the system's parameters changing rapidly with time. Talking about identification or estimation, a reliable control architecture should firstly estimate the parameters variations accurately in the presence of disturbances and secondly take a good control action. It has been shown that classical adaptive control methods do not operate well in this kind, giving rise to problems of instability as well as to large and oscillatory responses. Additionally, even when the control problem is solved efficiently there may be space for an improvement in the transient response of the system. The aforementioned difficulties lead researchers to the use of multiple models switching control.

Using the example of an LTI plant we will provide the basic steps in order to switch from the single adaptive model to the multiple models control scheme. The state space dynamic equation of an LTI plant S_p in its controller canonical form is given by:

$$S_p : \dot{x}_p(t) = A_p x_p(t) + b u(t) \tag{3.1}$$

where $x_p(.) \in R^n$, $u(.) \in R$, $A_p \in R^{n \times n}$ and $b \in R^n$. A_p and b can be defined as follows,

$$A_m = \begin{bmatrix} 0 & & & \\ 0 & & \mathbf{I}_{(n-1)} & \\ \vdots & & & \\ \alpha_{p1} & \alpha_{p2} & \cdots & \alpha_{pn} \end{bmatrix}, b = \begin{bmatrix} 0 \\ 0 \\ \vdots \\ 1 \end{bmatrix}$$

In this example, matrix A_p contains n unknown or uncertain parameters i.e. a_{p1}, a_{p2}, ..., a_{pn}. These n parameters has to be estimated in order to finally control the system, based on these estimations. It is supposed that the state $x_p(t)$ and the control signal $u(t)$ are accessible. The unknown parameters of A_p are assumed to be bounded with known bounds i.e. $\theta_p = \begin{bmatrix} a_{p1}, a_{p2}, \ldots, a_{pn} \end{bmatrix}^{\mathrm{T}} \in \Xi_\theta \; \theta_p : \underline{\theta}_p \leq \theta_p \leq \bar{\theta}_p$. The control signal has two objectives: (i) to stabilize the system and (ii) to force the system to follow a reference model given by the following equation,

$$S_m : \dot{x}_m(t) = A_m x_m(t) + br(t) \tag{3.2}$$

where $r(t) \in R$ is a known reference signal. The matrix A_m is given as follows,

$$A_p = \begin{bmatrix} 0 & & & \\ 0 & & \mathbf{I}_{(n-1)} & \\ \vdots & & & \\ \alpha_{m1} & \alpha_{m2} & \cdots & \alpha_{mn} \end{bmatrix}$$

According to the expression above, the main objective of the controller is to keep all the signals of the system bounded and also $\lim_{t \to \infty} [x_p(t) - x_m(t)] = 0$. The expression of the identification model is given as follows,

$$S_{id} : \dot{x}(t) = A_m x(t) + [A(t) - A_m]x_p(t) + bu(t) \tag{3.3}$$

where $A(t)$ is given as follows,

$$A_{(t)} = \begin{bmatrix} 0 & & & \\ 0 & & \mathbf{I}_{(n-1)} & \\ \vdots & & & \\ \alpha_1(t) & \alpha_2(t) & \cdots & \alpha_n(t) \end{bmatrix}$$

and $\theta^T(t) = [a_1(t), a_2(t), \ldots, a_n(t)]$ is the vector of the estimated plant parameters. We define $\tilde{\theta}(t) = \theta(t) - \theta_p$ as the parameter error and and $e(t) = x(t) - x_p(t)$ as the identification error. The time derivative of the identification error is given by the following equation,

$$\dot{e}(t) = A_m e(t) + b\tilde{\theta}^T(t)x_p(t) \tag{3.4}$$

The stability of the system and the boundedness will be assured by using the following Lyapunov function,

$$V(e, \tilde{\theta}) = e^T P e + \tilde{\theta}^T \tilde{\theta} \tag{3.5}$$

where P is a positive definite matrix obtained by the solution of the Lyapunov equation $A_m^T P + P A_m = -Q$, $Q = Q^T > 0$. The time derivative of (3.5) is given below,

$$\dot{V}(e, \tilde{\theta}) = -eQe + 2e^T P b\tilde{\theta}^T x_p + 2\tilde{\theta}^T \dot{\tilde{\theta}}. \tag{3.6}$$

The adaptive law for the estimates of the plant parameters is given as follows,

$$\dot{\theta}(t) = -x_p e^T(t) P b. \qquad (3.7)$$

Using (3.7) $\dot{V}(e, \tilde{\theta}) = -eQe \leq 0$ which means that the identification error $e(t)$, the parameter error $\tilde{\theta}(t)$, the parameters estimation vector $\theta(t)$ are bounded and the identifier is stable.

The second objective is to assure the stability of the plant and the boundedness of state vector $x_p(t)$ by choosing an appropriate control signal that takes into account the reference model (3.35). The state feedback control law is chosen as follows,

$$u(t) = k^T(t) x_p(t) + r(t) \qquad (3.8)$$

where $k(t) = \theta_m - \theta(t)$. Assuming that $e_c(t) = x_p(t) - x_m(t)$ is the control error, the time derivative of $e_c(t)$ is given by the following equation,

$$\dot{e}_c(t) = A_m e_c(t) - b\tilde{\theta}^T(t) x_p(t). \qquad (3.9)$$

In addition, if we use the direct adaptive control approach (3.9) can be written as

$$\dot{e}_c(t) = A_m e_c(t) + b\tilde{k}^T(t) x_p(t). \qquad (3.10)$$

where $\tilde{k}(t) = k(t) - k^\star$ and $k^\star = \theta_m - \theta_p$. Then the adaptive law for $k(t)$ is given as follows,

$$\dot{k}(t) = -x_p(t) e_c(t) P b \qquad (3.11)$$

According to the classical adaptive control theory it follows that $e_c \in L_2 \cap L_\infty$ and so the vectors $x_p(t)$ and $\dot{e}_c(t)$ are bounded. From Barbalat's lemma it follows that $\lim_{t\to\infty} e_c(t) = 0$ which means that the state $x_p(t)$ of the plant follows the state $x_m(t)$ of the reference model asymptotically. Also if the input $u(t)$ is persistently exciting then the estimated parameters will approximate the the real parameters of the plant as $t \to \infty$.

Theoretically, this method—and all the other single model adaptive methods that mentioned in the introduction—is very efficient with LTI systems but its performance depends mostly on the initialization of the identification model. If the initial estimation for the plant's parameters is far away from the real values there may be instability or bad transient response issues. These disadvantages are easily rebut by using multiple switching adaptive controllers along with their multiple identification models. In some cases to which we will refer in the sequence the multiple models may be either all fixed or fixed together with adaptive ones.

In Fig. 3.1, the general architecture of a multiple models switching control scheme is given. These models could be either fixed or adaptive. The description of this control scheme follows. The unknown plant model receives an input from

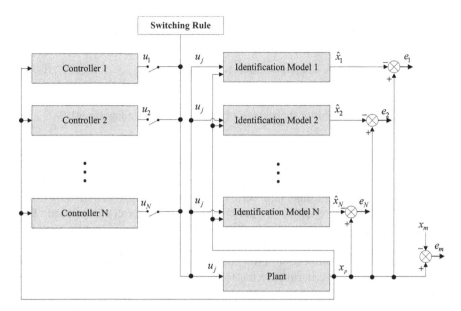

Fig. 3.1 The multiple models switching control architecture

one of the available controllers and produces an output which is equal to its state vector. The control objective is to make the plant's state track the reference model's state i.e. $e_m = x_p - x_m \to 0$. In order to achieve this objective we use N identification models $\{\mathcal{M}_k\}_{k=1}^N$ of the plant, which are operating in parallel. These N identification models are of the same architecture but they differ in their initial estimations for the plant's parameters. The initial parameters' estimates are distributed *uniformly* in space Ξ_θ. Every identification model produces an output $\hat{x}_k$, which is equal to its state vector and the identification error is given by $e_k = x - \hat{x}_k$. For every identification model $\mathcal{M}_k$, there is a corresponding fixed or adaptive controller $\mathscr{C}_k$ with an output u_k, with the controllers being parametrized using the certainty equivalence approach. By applying the controller $\mathscr{C}_k$ to the identification model $\mathcal{M}_k$, the output is given by a dynamical equation identical to that of the reference model (3.35). The objective of this architecture is to improve the performance of the system, and ensure its stability. This is performed by choosing at every time instant the appropriate controller according to a switching rule, which is based on a cost criterion J_k.

Utilizing the control architecture that described above with adaptive identification models we obtain N different identification models for the plant of the above example. These models are given as follows,

$$\mathcal{M}_k : \dot{\hat{x}}_k(t) = A_m \hat{x}_k(t) + [A_k(t) - A_m]x_p(t) + bu(t) \qquad (3.12)$$

where $k = 1, \ldots, N$ and $\hat{x}_k(t_0) = x_p(t_0)$. This means that all the identification models and the plant start from the same state value. The identification errors are

$e_k(t) = \hat{x}_k(t) - x_n(t)$ and the time derivative of these errors are given from the following equations,

$$\dot{e}_k(t) = A_m e_k(t)_b \tilde{\theta}_k^T(t) x_p(t). \tag{3.13}$$

where $\theta_k(t_0) = \theta_{k0}$ and $e_k(t_0) = 0$.

Using Lyapunov stability theory one obtains the weight adaptation law,

$$\dot{\theta}_k(t) = \dot{\tilde{\theta}}_k(t) = -x_p(t) e_k^T(t) Pb \tag{3.14}$$

Based on the estimations $\theta_k(t)$ of the plant parameters one can use a variety of methods to tune the controllers (e.g. certainty equivalence approach) and by using an appropriate cost criterion J_k the most suitable controller could be used. The general rule declares that the bast available controller is the controller with the minimum J_k at that time instant. There are may types of cost criteria but all of them contain the identification error in their equations. The most popular expressions of the cost criterion are the following,

1. $J_k(t) = e_k^2(t)$
2. $J_k(t) = \int_0^t e_k^2(r) \mathrm{d}r$
3. $J_k(t) = a e_k^2(t) + \beta \int_0^t e_k^2(r) \mathrm{d}r$
4. $J_k(t) = a e_k^2(t) + \beta \int_0^t e^{-\lambda(t-r)} e_k^2(r) \mathrm{d}r.$

It is obvious that the choice of the controller at any time instant and the performance of the system depends on the choice of the cost criterion. The first criterion chooses the controller according to the instant identification error, while the second criterion chooses the controller according to the long term identification error. The third criterion is a combination of the first and second criteria and finally the fourth criterion contains a forgetting factor λ. In order to avoid very fast or very slow controllers' switching we usually do not use the first and the second criterion. Also there are some other tools that are used in the switching procedure in order to ensure stability and reduce oscillatory responses. These are: a $T_{\min}$ waiting period between switches and an additive hysteresis h in the switching rule [10, 68].

In order to obtain a satisfactory transient response and mainly a better response than using a single model, it is very important to use the right number and type of identification models. The number of models depends on the size of the uncertainty region. Using a small number of models in a large region of uncertainty reduces the possibilities for a significant transient response improvement. Also the models should be uniformly distributed in the space of uncertainty.

3.2.2.2 Multiple Fixed Models Switching Control

As it was mentioned above, it is possible to use multiple fixed models in the control architecture which is depicted in Fig. 3.1 instead of using multiple adaptive models. In this case the control system becomes faster because the fixed models and controllers do not update their parameters at all. In this case, however, the accuracy of the control scheme is compromised. Also there is a condition that should be valid; At least one of the fixed models must be near the real plant in order to obtain a switching that results in stability. The controller that stabilizes this fixed model should stabilize the real plant too. It has been observed that the number of the fixed models increases exponentially with the dimension of the unknown vector and the size of uncertainty region. The crucial issue in this kind of control is to ensure the stability. It has been shown that stability is not assured for arbitrary switching schemes, since there may be controllers that destabilize the plant. In [10] the authors proved that when the waiting time $T_{\min}$ between switchings belongs to $(0, T_{\Xi})$ where T_{Ξ} is a positive number and when there is at least one model $\mathcal{M}_k$ with parameter error $\|\theta_k(t) - \theta_p\| < \mu_s(p, T_{\min})$, then all the signals in the overall system, as well as the cost criteria J_k, are uniformly bounded. It should be noted that T_{Ξ} depends only upon Ξ, and μ_s depends upon Ξ and a, β, λ form the aforementioned cost criteria.

In Nazrenda and Han [69] the authors proposed the redistribution of the fixed models in order to improve the performance. Redistribution is needed when the number of fixed models is small and so there is a need to change the initial location of the models, based on an observation of the system over a finite time interval. However the redistribution procedure makes the control algorithm more complex and less theoretically tractable.

3.2.2.3 A Combination of Fixed and Adaptive Multiple Models

Fixed and and adaptive identification models when used together in the framework of a multiple models architecture offer many advantages. Two are the main strategies of these hybrid schemes. These are:

(i)   the use of $N - s$ fixed models and s free adaptive models
(ii) the use of $N - 2$ fixed models, one free adaptive and one reinitialized adaptive model.

Based on these strategies, researchers have developed many other variations of these schemes. In both cases there is no need for using a large number of fixed models in order to assure stability and a satisfactory transient response. This is assured by using the adaptive models which are operating in parallel with the fixed models. Also, the requirement that the parameter error of at least one fixed model should be small enough is not essential because the identification errors of the adaptive models grow at a slower rate than the state of the system. Consequently there is no need to have a minimum number of fixed models in order to ensure stability and stability is independent of the parameters of the switching scheme. In the first scheme, the parameters of the adaptive models and their corresponding adaptive controllers are tuned by using the classic adaptive control theory techniques. The switching between the fixed controllers and the adaptive controllers is orchestrated by a switching rule and a cost criterion. In [10] it has been proved that if N_1 and $N_2 \geq 1$ fixed and free adaptive models respectively are used in a switching scheme with β, λ and $T_{\min}$, there exists a $T_{\Xi} > 0$ such that if $T_{\min} \in (0, T_{\Xi})$, then all signals in the overall system, as well as the cost criteria J_k are uniformly bounded. T_{Ξ} depends only upon Ξ.

In the second scheme the philosophy is the same but there is a substantial difference. Namely, the use of a reinitialized adaptive model. It is obvious from the aforementioned that the use of only free adaptive models may not be ideal especially when the initial parametric errors of the adaptive models are very large. Also when the plant is time-varying the free adaptive models have to change very frequently the direction of the update of their parameters. This problem has been partially solved by using the reinitialized adaptive model. This means that, at any time instant, if one of the fixed models is closest to the real plant, then the adaptive model reinitializes its parameter values and its state vector to that of the fixed model. Also, the free adaptive model is necessary to ensure the stability of the control scheme. Consequently, the second scheme ensures stability and a better transient response although is more complicated.

In the next subsection there is a presentation of the conventional methods that have been developed for the control of nonlinear systems using multiple models.

3.2.2.4 Nonlinear Systems Control Using Multiple Models

Nonlinear systems control using multiple models techniques is much more complicated than in the case of linear systems. In [12] the authors used multiple models in order to control parametric strict-feedback nonlinear systems of the following form,

$$\dot{x}_1 = x_2 + \varphi_1(x_1)^T \theta$$
$$\dot{x}_2 = x_3 + \varphi_2(x_1, x_2)^T \theta$$
$$\dot{x}_{n-1} = x_n + \varphi_{n-1}(x_1, x_2, \ldots, x_{n-1})^T \theta \qquad (3.15)$$
$$\vdots$$
$$\dot{x}_n = \beta(x)u + \varphi_n(x)^T \theta$$

where θ is a vector of unknown parameters, and β, φ_1, φ_2, ..., φ_n are smooth nonlinear functions. The usual methodologies used a backstepping adaptive controller for the system, some recursive equations and a positive definite control Lyapunov function which was tuned appropriately with an adaptive law for $\hat{\theta}(t)$ in order to result in a negative definite time derivative i.e. $\dot{V}(t) \leq 0$. In the aforementioned chapter this design was extended by allowing the parameter estimate to be reset instantaneously from $\hat{\theta}(t)$ to $\hat{\theta}(t^+)$ at any time instant. At these time instants the parameter adaptive law does not apply, and the control Lyapunov function may be discontinuous. The crucial issue in this chapter was to formulate reset conditions that could ensure that the control Lyapunov function would make a negative jump at the time of reset. By this way the stability is ensured and the convergence rate of the parameter estimator is getting faster. The role of multiple models here, is to provide the parameter vectors that give a negative jump to the Lyapunov functions and finally to select the one that gives the largest guaranteed decrease in $\Delta V(t)$ where $\Delta V(t) = V(x(t), \hat{\theta}(t^+)) - V(x(t), \hat{\theta}(t))$.

In [70] the authors apply the multiple models control architecture to nonlinear systems in the Brunovsky form:

$$\dot{x}_1 = x_2$$
$$\vdots \tag{3.16}$$
$$\dot{x}_{n-1} = x_n$$
$$\dot{x}_n = f^T(x_1, x_2, \ldots, x_n)\theta + u$$

or in a compact form,

$$\dot{x} = Ax + b(f^T(x)\theta + u) \tag{3.17}$$

where $\theta \in R^p$ is the unknown parameter vector, $x = (x_1, x_2, \ldots, x_n)^T$ is the state vector which can be measured, $f(x)$ is a known continuous vector field and $b = (0\ldots01)^T \in R^n$. The unknown parameters belong to a compact set Ξ and the objective is to ensure that the state $x(t)$ of the system (3.17) tracks the state of the following stable reference model,

$$\dot{x}_m = A_m x_m + br \tag{3.18}$$

There is a vector v such that $A_m = A + bv^T$ and if $u = v + v^T x$, the kth identification model can be described by the following equation,

$$\dot{x}_k = A_m x_k + b(f^T(x)\theta_k + v) \tag{3.19}$$

The identification error for model $\mathcal{M}_k$ is $e_k(t) = x_k(t) - x(t)$ and its cost criterion is given as follows,

$$J_k(t) = a_1 \|e_k(t)\|^2 + a_2 \int_0^t \|e_k(r)\|^2 dr \tag{3.20}$$

where $a_1, a_2 \geq 0$. The model with the minimum J_k defines the controller for the system for that time instant. Supposing that $J_j(t) = \min_{k \in N}\{J_k(t)\}$, the control signal is given below,

$$v(t) = f^T(x)\theta_j + r(t). \tag{3.21}$$

By using Lyapunov theory, the authors prove that the plant (3.17) with the identification models (3.19), the cost criterion (3.20), a $T_{\min}$ between the switchings and an adaptive law given form (3.22),

$$\dot{\theta}_k = -(e_k)^T Pbf(x) \tag{3.22}$$

will track the desired state $x_m(t)$ asymptotically with zero error i.e. $\lim_{x\to\infty} \|x(t) - x_m(t)\| = 0$ and all the signals in the system will remain bounded.

In case $N - 1$ fixed models and one free adaptive model are used in the control architecture, the proof is similar to the case of all adaptive models. We have to note here that the cost criterion of the free adaptive model is bounded while those of the fixed models grow in an unbounded fashion. Consequently, there exists a finite time T such that the system switches to the controller corresponding to the adaptive model and stays there for $t \geq T$, unless the plant is time-varying. In that case the procedure will start form the beginning. In case, a reinitialized adaptive model is added in the architecture, the stability and boundedness proof is much the same with the aforementioned.

In [71], the authors use multiple models in the model reference adaptive control framework in order to control a class of minimum phase nonlinear systems with large parametric uncertainties and improve its transient response. The proposed approach consists of $N + 1$ models. N models are fixed and one model is the free running adaptive model. A parameter reset condition is defined such that the parameter update rule of the adaptive model resets itself to the parameter set which leads to a negative jump for the corresponding Lyapunov function. This negative jump is valid over a predefined time period called the permissible switching time.

Ye [15] uses N adaptive identification models to control nonlinear systems in parametric strict feedback form. The identification models are uniformly distributed in the compact set Ξ of uncertainty and the switching criterion contains a $T_{\min}$ between switchings and an additive hysteresis h. This hysteresis is added to the cost criterion value of the model with the minimum cost criterion and if this sum is less than the value of the cost criterion of the current (dominant) model then a switch happens. Also, the author provides an analysis provin that after a finite time, only one of the adaptive controllers will dominate to the system control and the switchings will be terminated.

In the following sections some of the latest advances in intelligent multiple models based adaptive switching control are given.

3.3 Multiple Models Switching Control and Neural Networks

Neural networks have been playing a key role in the adaptive control field since 1990s and researchers used them in the framework of multiple models control also. In this section, based on a fundamental work, the role of neural networks in that kind of architectures is presented.

3.3.1 Multiple Models Adaptive Control for SISO and MIMO Discrete-Time Nonlinear Systems Using Neural Networks

The motivation for using neural networks in a switching scheme comes from the multiple models control philosophy. To be more specific, multiple models offer the chance to the engineer to use different kind of models together, in order to exploit the advantages of each model and improve the performance of the nonlinear system. In [20] the authors proposed a switching scheme between a linear robust adaptive controller and a neural network based nonlinear adaptive controller in order to control a class of nonlinear discrete-time dynamical systems and to ensure the boundedness of all signals. There is a neural network model that is used to approximate complex nonlinear dynamic systems whose mathematical models are not available or are very complicated. Based on this estimation another neural network plays the role of the adaptive controller. Also, there are a linear model and a linear robust controller which have two objectives: to provide stability and a lower bound on the performance of the system. In the beginning, the linear robust controller takes over the control of the system. As the time passes by and the neural network model keeps training itself, the switching system converges to the neural controller and finally the performance is improved. The key issue in this architecture is to choose an appropriate switching rule that will provide the best possible results. The basic points of this architecture are given below.

Let the following SISO discrete-time nonlinear system [20],

$$\begin{aligned} x(k+1) &= F(x(k), u(k)) \\ y(k) &= H(x(k)) \end{aligned} \tag{3.23}$$

where $u(k)$, $y(k) \in R$, $x(k) \in R_n$ and F, H are smooth nonlinear functions such that the origin is an equilibrium state. A linearization around the equilibrium state gives the following description for (3.23),

$$x(k+1) = Ax(k) + bu(k) + \bar{f}(x(k), u(k))$$
$$y(k) = cx(k) + \bar{h}(x(k))$$

$$(3.24)$$

where $A \in R^{n \times n}$, b, $c^T \in R^n$ and the triple (c, A, b) represents the linearization of (3.23). The nonlinear functions $\bar{f}, \bar{h}$ belong to the class of high order functions and are obtained by subtracting the linear terms form the functions F, H. For an observable system of order n the state $x(k+1)$ can be described by the following equation,

$$y(k+d) = \theta^T \omega(k) + f(\omega(k))$$

$$(3.25)$$

where $\omega(k) = [y(k), ..., y(k-n+1), u(k), ..., u(k-n+1)]^T$ is the regression vector, $\theta = [a_0, ..., a_{n-1}, b_0, ..., b_{n-1}]^T$, $b_0 \geq b_{min} > 0$ is the linear unknown parameter vector, and f is a smooth unknown nonlinear function that consists of high order terms of $\omega(k)$ locally. The objective of the adaptive controller is to generate a bounded control signal $u(k)$ such that the output of the system asymptotically approaches a pre-specified desired signal $y^\star(k)$ i.e. $\lim_{k \to \infty} |y(k) - y^\star(k)| = 0$.

In case the nonlinearity f is small, the linear robust control techniques are the appropriate tools to handle it as a bounded disturbance, but when the nonlinearity is not so small other methods must be used in order to handle it. Here, neural networks are used for that reason. It must be noted that the nonlinearity f is globally bounded, the system has a globally uniformly asymptotically stable zero dynamics and the linear parameter vector lies in a compact region Ξ.

The switching architecture is depicted in Fig. 3.2. The control scheme uses two models, one linear and one neural network model along with their corresponding controllers. In general cases the number of models and controllers could be larger than two. Here, $\mathcal{M}_1$, $\mathcal{M}_2$ are the two models which try to estimate the plant's parameters. Their parameters are updated by using two inputs; the control input and the output of the plant. The cost criteria J_1, J_2 define which controller will dominate at every time instant. The model $\mathcal{M}_j$ with the minimum J_k, $k = 1, 2$ will be chosen to generate a certainty equivalence control signal $u(t)$ for the plant. This means that the output of $\mathcal{M}_j$ will be identical to the desired reference value. The choice of the appropriate switching scheme is very important for the success of the control system.

The linear adaptive identification model can be defined as,

$$\hat{y}_1(k+1) = \hat{\theta}_1^T(k)\omega(k)$$

$$(3.26)$$

and the adaptive law for $\hat{\theta}_1(k)$ is given as follows,

$$\hat{\theta}_1(k) = \hat{\theta}_1(k-1) - \frac{a(k)e_1(k)\omega(k-1)}{1 + |\omega(k-1)|^2}$$

$$(3.27)$$

Fig. 3.2 The multiple models switching control architecture

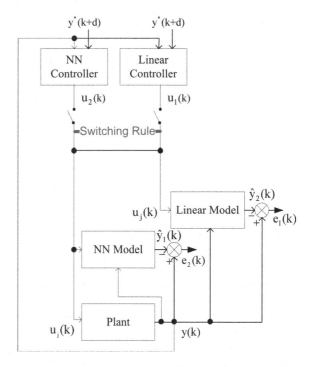

where $e_1(k) = \hat{y}_1(k) - y(k)$, $a(k) = \begin{cases} 1 & if \ |e_1(k)| > 2D\Delta \\ 0 & otherwise \end{cases}$, Δ is the bound of nonlinearity and $\hat{b}_0^{(1)}(k)$ in $\hat{\theta}_1(k)$ is constrained to be greater than a $b_{min} > 0$. The corresponding controller is given as follows,

$$u_1(k) = \frac{1}{\hat{b}_0^{(1)}(k)} \left(y^*(k+1) - \hat{\bar{\theta}}_1(k)\bar{\omega}(k) \right) \tag{3.28}$$

where $\bar{\theta} = [a_0, \ldots, a_{n-1}, b_1, \ldots, b_{n-1}]^T$ and $\bar{\omega}(k) = [y(k), \ldots, y(k-n+1), u(-1.k), \ldots, u(k-n+1)]^T$. The model M_2 can be defined as follows,

$$\hat{y}_2(k+1) = \hat{\theta}_2^T(k)\omega(k) + \hat{f}(\bar{\omega}(k), W(k)) \tag{3.29}$$

where $\hat{f}(.)$ is a bounded continuous nonlinear function parameterized by a neural network weights vector $W(k)$. Its parameters $\hat{\theta}_2(k)$ or $W(k)$ are updated with no restriction except the fact that they always lie inside a pre-defined compact region Ψ i.e.

$$\hat{\theta}_2(k), \hat{W}(k) \in \Psi, \forall k \tag{3.30}$$

The corresponding controller is given as follows,

$$u_2(k) = \frac{1}{\hat{b}_0^{(2)}(k)} \left(y^*(k+1) - \hat{\theta}_2^T(k) - \hat{f}(\bar{\omega}(k), W(k)) \right). \tag{3.31}$$

The cost criterion is given by the following equation,

$$J_j(k) = \sum_{l=1}^{k} \frac{a_j(l)(e_j^2(l) - 4\Delta^2)}{2(1 + |(\omega(l-1))^2|)} + c \sum_{l=k-N+1}^{k} (1 - a_j(l))e_j^2(l), j = 1, 2 \tag{3.32}$$

where $e_j(k) = \hat{y}_j(k) - y(k)$, $a_j(k) = \begin{cases} 1 & if\ |e_j(k)| > 2\Delta \\ 0 & otherwise \end{cases}$, N is an integer and $c \geq 0$ is a constant.

The switching rule turns the switch to the controller that corresponds to the model with the minimum cost criterion J.

The authors proved that by using the identification models (3.26), (3.29), with their adaptive laws (3.27), (3.30) and the cost criterion (3.32) with its switching rule, all the signals in the systems are bounded and if the parameter adjustment mechanism of the neural network reduces the identification error, then the tracking error can go to zero.

The simulations showed that when using the switching scheme with the linear and neural network controller the results are much better than using only the linear or the neural network controller alone.

Fu and Chai [22] extended the work of Chen and Narendra [20] to MIMO systems, relaxed the assumption for a globally bounded high-order nonlinearity of the system and provided a tracking error analysis. The results are much the same with [20] and the simulations in a ball mill pulverizing system showed that the proposed methodology is robust to external disturbances and the transient response is very satisfactory.

Although there are not many works on multiple models neurocontrol, from the aforementioned it is obvious that neurocontrol can affect positively the performance of a nonlinear system when it is used in the framework of multiple models. In the next section, the latest trends in multiple models adaptive fuzzy switching control are described analytically.

3.4 Multiple Models Switching Control and Fuzzy Systems

In this section, the contribution of Takagi-Sugeno (T-S) fuzzy systems to the field of multiple models switching control is given analytically. Two different architectures are described and all their advantages and disadvantages are presented.

3.4.1 The Role of T-S Fuzzy Systems

Control of nonlinear and unknown or highly uncertain systems, is a very challenging and difficult task for the engineers due to the fact that nonlinear control techniques are not systematically developed and the unknown parameters impose a negative impact to the performance of these systems. Concerning the nonlinearity problem, fuzzy logic techniques have been shown to be an adequate and very useful tool.

During the last three decades there have been remarkable efforts from researchers in the field of fuzzy control systems. In [72] a fuzzy model (T-S model) that uses a set of fuzzy rules to describe a nonlinear system in terms of some local linear subsystems was suggested. These linear models—which are found in the consequent part of the rules—are fuzzy blended and the model of the nonlinear system is obtained. The main advantage of T-S models is that they offer to the engineer the possibility to utilize linear control techniques in order to control the global nonlinear system. Another advantage of T-S models is that they have the so-called "universal function approximation" property, in the sense that they are able to approximate every smooth nonlinear function to any degree of accuracy in a convex compact region.

T-S fuzzy systems are suitable to be combined with adaptive control techniques due to the fact that they consist of fuzzy blended T-S models which contain linear dynamic equations in the consequent parts of the fuzzy rules that describe them.

The combination of T-S fuzzy systems and adaptive control offers control schemes that can be used to control unknown or uncertain plants and many works have been published since the mid-1990s. But, why this combination do not always give good results and which is the role of multiple models in that case?

3.4.2 Multiple T-S Fuzzy Models for More Reliable Control Systems

When a single T-S adaptive fuzzy model along with its adaptive fuzzy controller in an indirect scheme is used to control an unknown or highly uncertain nonlinear system there may be some stability or transient response performance problems. This could happen if the initial estimations for the original parameters of the plant are highly inaccurate or if the parameters of the plant are changing rapidly and discontinuously. In [23–25] the authors proposed the T-S multiple models based adaptive switching control (TSMMASC) architecture which is capable of facing successfully this kind of problems. The idea was to use multiple fuzzy T-S models along with their corresponding controllers in an indirect adaptive switching control scheme. By this way all the advantages of fuzzy modeling are combined with the advantages of multiple models in adaptive control and the results are very promising. In the following sections an analytical description of the architecture is given and some paradigms confirm the effectiveness of this intelligent control approach.

3.4.3 Problem Statement and Single Identification Model

In this section the mathematical expressions of the plant to be controlled and its identification model are given. Let a continuous-time nonlinear unknown system described by using the T-S method. The fuzzy rules for a nth order plant are of the following form:

$$Rule\ i : \text{IF } x_1(t) \text{ is } M_1^i \text{ and}\dots\text{and } x_n(t) \text{ is } M_n^i$$
$$\text{THEN } \dot{x}(t) = A_i x(t) + B_i u(t)$$

where $i = 1, \ldots, l$ is the number of fuzzy rules, M_p^i, $p = 1, \ldots, n$ are the fuzzy sets, $x = [x_1, x_2, \ldots, x_n]^T \in R^n$, $u \in R$ are the state vector and control input of the system respectively and $A_i^{n \times n}$, $B_i^{n \times 1}$ are the state and input matrices respectively, which are considered to be unknown.

The matrices A_i, B_i are of the following form:

$$A_i = \begin{bmatrix} 0 & & & \\ 0 & & \mathbf{I}_{n-1} & \\ \vdots & & & \\ \alpha_n^i & \alpha_{n-1}^i & \cdots & \alpha_1^i \end{bmatrix}, \ B_i = \begin{bmatrix} 0 \\ 0 \\ \vdots \\ b^i \end{bmatrix}$$

where $\mathbf{I}_{(n-1)}$ is an $(n-1) \times (n-1)$ identity matrix.

Given a pair of $x(t)$, $u(t)$, the final form of the fuzzy system is given by the following equation:

$$\dot{x}(t) = \frac{\sum_{i=1}^{l} h_i(x(t))(A_i x(t) + B_i u(t))}{\sum_{i=1}^{l} h_i(x(t))} \tag{3.33}$$

where $h_i(x(t)) = \prod_{p=1}^{n} M_p^i(x_p(t)) \geq 0$ and $M_p^i(x_p(t))$ is the grade of membership of $x_p(t)$ in M_p^i for all $i = 1, \ldots, l$ and $p = 1, \ldots, n$.

The state space parametric model (SSPM) [73] expression of (3.33) is given by the following equation:

$$\dot{x}(t) = A_d x(t) + \frac{\sum_{i=1}^{l} h_i(x(t))((A_i - A_d)x(t) + B_i u(t))}{\sum_{i=1}^{l} h_i(x(t))} \tag{3.34}$$

The matrix A_d is stable and corresponds to the state matrix of an arbitrary reference model which is described by the following equation:

$$\dot{x}_m = A_d x_m \tag{3.35}$$

where $x_m \in R^n$ is the state vector of the desired reference model. According to the problem statement, the parameters matrices A_i, B_i are unknown and an estimation model must be used in order to design the control signal based on the certainty equivalence approach. Utilizing the series parallel model (SPM) [73] the estimation model can be formed as:

$$\dot{\hat{x}}(t) = A_d \hat{x}(t) + \frac{\sum_{i=1}^{l} h_i(x(t))((\hat{A}_i - A_d)x(t) + \hat{B}_i u(t))}{\sum_{i=1}^{l} h_i(x(t))} \tag{3.36}$$

where the symbol "∧" denotes the estimated values for the parameters of the real plant. In the next section, the TSMMASC architecture [24] along with a switching rule and a cost criterion which are used together are presented.

3.4.4 Architecture

The TSMMASC architecture is depicted in Fig. 3.3 and is described as follows; The unknown T-S plant model receives an input from one of the available fuzzy adaptive controllers and produces an output which is identical to its state vector. The control objective is to make the plant's state track the reference model's state i.e. $e_m = x - x_m \to 0$. In order to achieve this objective, N T-S identification models $\{\mathcal{M}_k\}_{k=1}^{N}$ of the plant are used which are operating in parallel. Every linear submodel of each T-S model rule is expressed by using the SPM formulation (3.36). These N T-S models are of the same architecture concerning the number of rules, the membership functions and the premise variables but they differ in their initial estimations for the plant's parameters. More precisely, the uncertain parameters of A_i, B_i are denoted as $\mathcal{E}_{AB} \in \Xi \subset R^s$, where Ξ is a compact space indicating the region of all the possible parameters values combinations and s is equal to the number of the unknown parameters. The initial parameters estimates are distributed *uniformly* in space Ξ. Every T-S model produces an output $\hat{x}_k$, which is equal to its state vector and the identification error is given by

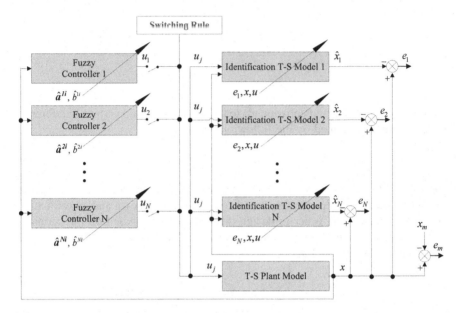

Fig. 3.3 The TSMMASC architecture

$e_k = x - \hat{x}_k$. For every identification model $\mathcal{M}_k$, there is a corresponding feedback
linearization adaptive controller $\mathcal{C}_k$ with an output u_k, and all the controllers are
updated in an indirect way, using the certainty equivalence approach. By applying
the controller $\mathcal{C}_k$ to the T-S model $\mathcal{M}_k$, the output is given by a dynamical
equation identical to that of the reference model (3.35). The objective of the
TSMMASC architecture is to improve the performance of the system, and ensure
its stability. This is performed by choosing at every time instant the appropriate
controller according to a switching rule, which is based on a cost criterion J_k.

3.4.5 Switching Rule and Cost Criterion

In this subsection the description of the switching scheme which orchestrates the
suitable controller selection is given. A switching rule selects at every time instant
the most appropriate controller for the plant. This rule is based on some indices J_k
which are relevant to the model identification errors $e_k = x - \hat{x}_k$, while the performance of the identification T-S models is evaluated in parallel at every time
instant. An additive hysteresis constant h [6, 68], and a $T_{\min}$ [9] are also used. The
switching rule description follows.

If $J_f(t) = \min_{k \in \Lambda}\{J_k(t)\}$, $\Lambda = \{1, \ldots, N\}$, and $J_f(t) + h \leq J_{cr}(t)$ is valid in the
time interval $[t, t + T_{\min}]$, where $J_{cr}(t)$ is the index of the current active T-S model
$\mathcal{M}_{cr}$, then the controller $\mathcal{C}_f$ is chosen and is tuned utilizing the certainty equivalence approach from the estimations of the corresponding T-S model $\mathcal{M}_f$.

If $J_f(t) + h > J_{cr}(t)$ is valid at any time instant of the time interval $[t, t + T_{\min}]$, then the controller $\mathscr{C}_{cr}$ remains active, meaning that it is the ideal controller for that time interval. The above procedure is repeated at every step. It has to be noted that the signal of the dominant controller $\mathscr{C}_f$ or $\mathscr{C}_{cr}$, is used to control the plant as well as all the other T-S identification models. If $J_f(t)$ is not unique (i.e. there are two or more models with equal minimums), the choice of the dominant controller is made arbitrarily among these models.

The cost criterion is of the following form:

$$J_k(t) = a_c e_k^2(t) + b_c \int_0^t e_k^2(r)dr \qquad (3.37)$$

where a_c, b_c are design parameters. If $J_k(t) = e_k^2(t)$, possible parameters changes are detected very quickly but the switching between the controllers may be very rapid and the chattering effect may lead to poor control. On the other hand, if $J_k(t) = \int_0^t e_k^2(r)dr$, the parameters changes are not detected on time and the switching between the controllers may be infrequent. Consequently the criterion (3.37), which embodies both instantaneous and past measures, may lead to a smooth and satisfactory control signal in case the parameters of the plant are time-varying. A very significant remark follows about the role of $T_{\min}$.

Remark 1 In order to avoid switching at very high frequencies, a time interval $T_{\min}$ is allowed to elapse between the switchings so that $T_{i+1} - T_i \geq T_{\min}$, $\forall i$ where $\{T_i\}_{i=0}^\infty$ is a switching sequence with $T_0 = 0$ and $T_i < T_{i+1}$, $\forall i$. The choice of an arbitrarily small $T_{\min}$ leads in a globally stable system [9].

3.4.6 T-S Identification Models, Controller Design and Next Best Controller Logic

In this section, the appropriate mathematical expression for the T-S identification models is presented, the fuzzy controller is given and finally the next best controller logic (NBCL) is described.

3.4.6.1 T-S Identification Models

The TSMMASC approach makes use of N architecturally identical T-S models with different initializations concerning the parameters estimations. Every linear submodel in the consequent part of every T-S model rule is expressed by using the SPM formulation. The fuzzy rules that describe every T-S model $\mathscr{M}_k$ are of the following form:

T S Identification Model $\mathcal{M}_k$

Rule i : IF $x_1(t)$ is M_1^{ki} and $x_2(t)$ is M_2^{ki} and...and $x_n(t)$ is M_n^{ki}

THEN $\dot{\hat{x}}_k(t) = A_d\hat{x}_k(t) + (\hat{A}_{ki} - A_d)x(t) + \hat{B}_{ki}u(t)$

where $k \in \Lambda = \{1, ..., N\}$ and $i = 1, ..., l$. It should be noted that the number N of T-S identification models is independent of the number n of the state variables. The final form of every T-S model is inferred by a fuzzy blending of the linear subsystems and is given by the following equation:

$$\dot{\hat{x}}_k(t) = A_d\hat{x}_k(t) + \frac{\sum_{i=1}^{l} h_{ki}(x)((\hat{A}_{ki} - A_d)x(t) + \hat{B}_{ki}u(t))}{\sum_{i=1}^{l} h_{ki}(x)} \qquad (3.38)$$

where $h_{ki}(x) = \prod_{p=1}^{n} M_p^{ki}(x_p(t)) \geq 0$ and $M_p^{ki}(x_p(t))$ is the grade of membership of $x_p(t)$ in M_p^{ki}, $k \in \Lambda$, $i = 1, ..., l$ and $p = 1, ..., n$. Also, $M_p^{ki}(x_p(t)) = M_p^i(x_p(t))$ and $h_{ki}(x) = h_i(x)$ for all k, i, p. The matrices of all the T-S models are of the following form:

$$\hat{A}_{ki} = \begin{bmatrix} 0 & & & \\ 0 & & \mathbf{I_{(n-1)}} & \\ \vdots & & & \\ \hat{a}_n^{ki} & \hat{a}_{n-1}^{ki} & \cdots & \hat{a}_1^{ki} \end{bmatrix}, \quad \hat{B}_{ki} = \begin{bmatrix} 0 \\ 0 \\ \vdots \\ \hat{b}^{ki} \end{bmatrix}$$

3.4.6.2 Controller Design

Using the feedback linearization technique [74], and denoting the dominant controller as $\mathscr{C}_j$, the control signal at every time instant can be described by the following equation:

$$u(t) = u_j(t) = \frac{\sum_{i=1}^{l} h_{ji}(x)(\mathbf{a}^d - \hat{\mathbf{a}}^{ji})^T x(t)}{\sum_{i=1}^{l} h_{ji}(x)\hat{b}^{ji}} \qquad (3.39)$$

where $(\hat{\mathbf{a}}^{ji})^T = [\hat{a}_n^{ji} \ \hat{a}_{n-1}^{ji} \ \cdots \hat{a}_2^{ji} \ \hat{a}_1^{ji}]$ and $(\mathbf{a}^d)^T = [a_n^d \ a_{n-1}^d \ \cdots a_2^d \ a_1^d]$ and are the nth rows of the estimated state and reference matrices respectively.

Applying the control input $u(t)$ given by (3.39) to the dominant T-S model $\mathcal{M}_j$ and noting that $h_{ji}(x) = h_{ki}(x) = h_i$, it follows,

$$\dot{\hat{x}}_j(t) = A_d\hat{x}_j(t) + \frac{1}{\sum\limits_{i=1}^{l} h_i}\left(\sum_{i=1}^{l} h_i\left((\hat{A}_{ji} - A_d)x(t) + \hat{B}_{ji}\frac{\sum\limits_{i=1}^{l} h_i(\mathbf{a}^d - \hat{\mathbf{a}}^{ji})^T x(t)}{\sum\limits_{i=1}^{l} h_i \hat{b}^{ji}}\right)\right)$$

$$= A_d\hat{x}_j(t) + \frac{1}{\sum\limits_{i=1}^{l} h_i}\left\{\begin{bmatrix} 0 & & & \\ 0 & & \mathbf{0}_{(n-1)} & \\ \vdots & & & \\ \sum\limits_{i=1}^{l} h_i(\hat{a}_n^{ji} - a_n^d) & \sum\limits_{i=1}^{l} h_i(\hat{a}_{n-1}^{ji} - a_{n-1}^d) & \cdots & \sum\limits_{i=1}^{l} h_i(\hat{a}_1^{ji} - a_1^d) \end{bmatrix}\right]$$

$$+ \begin{bmatrix} 0 \\ 0 \\ \vdots \\ 1 \end{bmatrix}\begin{bmatrix} \sum\limits_{i=1}^{l} h_i(a_n^d - \hat{a}_n^{ji}) & \sum\limits_{i=1}^{l} h_i(a_{n-1}^d - \hat{a}_{n-1}^{ji}) & \cdots & \sum\limits_{i=1}^{l} h_i(a_1^d - \hat{a}_1^{ji}) \end{bmatrix}\right\}x(t).$$

where $\mathbf{0}_{(n-1)}$ is a $(n-1) \times (n-1)$ zero matrix.

It is obvious that:It can be noticed that when the input $u_j(t)$ is applied to the corresponding estimated plant $\mathcal{M}_j$, this plant is linearized and it is asymptotically stable, with an identical behavior to that of the desired reference model (3.35).

$$\dot{\hat{x}}_j(t) = A_d\hat{x}_j(t) \tag{3.40}$$

3.4.6.3 The Next Best Controller Logic

The NBCL is derived from the necessity to provide a feasible control signal during the control procedure of the plant. When a single adaptive feedback linearization controller of type (3.39) is used, there is a possibility to produce a zero control signal although $x(t) \neq 0$. This is the case when at any time instant the following equality holds for $j = 1$ (i.e. a single T-S identification model):

$$\sum_{i=1}^{l} h_{ji}(x)(\mathbf{a}^d - \hat{\mathbf{a}}^{ji})^T x(t) = 0 \tag{3.41}$$

Then, from (3.39) it holds that $u(t) = u_j(t) = 0$. Although the estimated parameters $\hat{\mathbf{a}}^{ji}$ will change their values at the next step, the zero control signal may lead to instabilities or poor performance of the plant. In case multiple models are used for the control of the plant, there is also a possibility for a zero control signal when (3.41) holds for the winning controller $\mathcal{C}_j$. In order to avoid this zero control signal, the following criterion imposed in the switching rule:

If C_j is the dominant controller and $\sum_{i=1}^{l} h_{ji}(x)(\mathbf{a}^d - \hat{\mathbf{a}}^{ji})^T x(t) = 0$ and $x(t) \neq 0$

Then $u(t) = u_{j(nbc)}(t)$, where $j(nbc) = \arg \min_{k \in \wedge, \, k \neq j} \{J_k(t)\}$.

This criterion forces the switching rule to choose the best alternative controller which will not provide a zero signal for that time instant.

3.4.7 Stability, Adaptive Law, Convergence and Computational Cost

The stability analysis and the adaptive law of the TSMMASC architecture along with a discussion on the convergence and the computational cost of the switching algorithm are given in this subsection.

3.4.7.1 Stability and Adaptive Laws

The stability in this architecture depends on the identification error of every T-S model which is defined as follows,

$$e_k = x - \hat{x}_k \tag{3.42}$$

The error e_k corresponds to the difference between the state of the plant and the state of the T-S model $\mathcal{M}_k$. The derivative of the identification error is given by the following equation:

$$\dot{e}_k = A_d e_k - \frac{\sum_{i=1}^{l} h_i \left[\mathbf{0}_{n \times (n-1)} \tilde{\mathbf{a}}^{ki} \right]^T}{\sum_{i=1}^{l} h_i} x - \frac{\sum_{i=1}^{l} h_i \left[0 \; 0 \; \cdots \; \tilde{b}^{ki} \right]^T}{\sum_{i=1}^{l} h_i} u \tag{3.43}$$

where $\tilde{a}_p^{ki} = \hat{a}_p^{ki} - a_p^i$, $\tilde{b}^{ki} = \hat{b}^{ki} - b^i$, $\tilde{\mathbf{a}}^{ki} = \left[\tilde{a}_n^{ki} \; \tilde{a}_{n-1}^{ki} \; \cdots \; \tilde{a}_1^{ki} \right]^T$ is a $n \times 1$ vector, $\mathbf{0}_{n \times (n-1)}$ is a zero matrix of dimension $n \times (n-1)$ and $p = 1, \ldots, n$.
    Lyapunov function candidates are given as follows,

$$V_k(e_k, \tilde{\mathbf{a}}^{ki}, \tilde{b}^{ki}) = e_k^T P_k e_k + \sum_{i=1}^{l} \frac{(\tilde{\mathbf{a}}^{ki})^T \tilde{\mathbf{a}}^{ki}}{r_1^{ki}} + \sum_{i=1}^{l} \frac{(\tilde{b}^{ki})^2}{r_2^{ki}} \tag{3.44}$$

where $r_1^{ki}, r_2^{ki} > 0$ are the learning rate constants, $V_k \geq 0$, and $P_k = P_k^T$ is the solution of the Lyapunov equation:

$$A_d^T P_k + P_k A_d = -Q_k \tag{3.45}$$

for known $Q_k > 0$.

The objective in this case is to ensure that

$$\dot{V}_k \leq 0 \tag{3.46}$$

After some mathematical analysis the adaptive law for the estimates of the multiple T-S models takes the following form,

$$(\dot{\hat{a}}^{ki})^T = r_1^{ki} \frac{h_i}{\sum\limits_{i=1}^{l} h_i} P_{ks}^T e_k x^T,$$

$$\dot{\hat{b}}^{ki} = \begin{cases} r_2^{ki} \dfrac{h_i}{\sum\limits_{i=1}^{l} h_i} P_{ks}^T e_k u, & \begin{cases} \text{if } |\hat{b}^{ki}| > b_0^i \text{ or} \\ \text{if } |\hat{b}^{ki}| = b_0^i \text{ and } P_{ks}^T e_k u \, sgn(b_i) \geq 0, \text{for all } i, k. \end{cases} \\ 0, & \text{if } |\hat{b}^{ki}| = b_0^i \text{ and } P_{ks}^T e_k u \, sgn(b_i) < 0, \text{for all } i, k. \end{cases} \tag{3.47}$$

where $P_{ks} \in R^{n \times 1}$ is the nth column of P_k. It is obvious that the adaptive law (3.47) contains a restriction for the update of $\dot{\hat{b}}^{ki}$. This restriction comes from the following assumption; the sign of b^i and a lower bound $b_0^i > 0$ for $|b^i|$ are known for all $i = 1, \ldots, l$. Consequently, the term $1/\hat{b}^{ki}$ in the control signal expression is bounded because the term $\hat{b}^{ki}$ never goes to zero or through zero. The initialization of the $\hat{b}^{ki}$ for every one of the N T-S models is chosen so that the following inequality holds:

$$\hat{b}^{ki}(0) \, sgn(b^i) \geq b_0^i \tag{3.48}$$

Considering the adaptive law (3.47), the time derivative of V_k takes the following form:

$$\dot{V}_k = \begin{cases} -e_k^T Q_k e_k, & \begin{cases} \text{if } |\hat{b}^{ki}| > b_0^i \text{ or} \\ \text{if } |\hat{b}^{ki}| = b_0^i \text{ and } P_{ks}^T e_k u \, sgn(b_i) \geq 0, \text{for all } i, k. \end{cases} \\ -e_k^T Q_k e_k - 2 \dfrac{\sum\limits_{i=1}^{l} h_i \tilde{b}^{ki} P_{ks}^T e_k u}{\sum\limits_{i=1}^{l} h_i}, & \text{if } |\hat{b}^{ki}| = b_0^i \text{ and } P_{ks}^T e_k u \, sgn(b_i) < 0, \text{ for all } i, k. \end{cases} \tag{3.49}$$

Using the above adaptive law the authors proved the following theorem,

Theorem 1

Given a plant model (3.33) with the control law (3.39), the adaptive law (3.47) and a reference model with state matrix A_d, the TSMMASC approach guarantees that for all $i = 1,...,l$ and $j, k \in \Lambda$:

1. $\hat{\mathbf{a}}^{ki}$, $\hat{b}^{ki}$, $1/\hat{b}^{ki}$, $e_k(t)$ *are bounbed,*
2. $\left[e_k(t), \dot{\hat{\mathbf{a}}}^{ji}(t), \dot{\hat{b}}^{ji}(t), e_m(t) \right] \to 0$ *as $t \to \infty$.*

The proof of this Theorem is given in [24].

The TSMMASC startegy can be delineated as follows:

Step 1. Represent the nonlinear plant using a T-S fuzzy model (3.33).

Step 2. Based on the representation of Step 1 and utilizing the SPM expression, construct N T-S identification models $\{\mathcal{M}_k\}_{k=1}^{N}$.

Step 3. For every identification model $\mathcal{M}_k$, construct a controller $\mathcal{C}_k$ based on a feedback linearization technique so that when the control signal u_k is applied in model $\mathcal{M}_k$, the dynamic equation of the specified reference model (3.35) is obtained.

Step 4. Define a cost criterion $J_k(t)$ of the form (3.37), a hysteresis h, a $T_{\min}$ and a switching rule for the controllers.

Step 5. For $t = 0$, choose arbitrarily the signal from one of the available controllers. For $t > 0$, repeat Steps 6–8 at each sample instant:

Step 6. Calculate all the identification errors e_k and use the adaptive law (3.47) to update the parameters estimations for every T-S model.

Step 7. Utilizing the updated parameters form Step 6, update the controllers $\mathcal{C}_k$.

Step 8. Apply the appropriate controller $\mathcal{C}_j$, based on criterion (3.37) and NBCL, to the plant and to all the T-S identification models.

3.4.7.2 Comments on Convergence and Computational Cost of the Switching Algorithm

Three are the main tools of the switching algorithm that are used in this control scheme. These are the cost criterion (3.37) which embodies past measures of the squared identification errors, the constant hysteresis h and a $T_{\min}$ between the successive switchings. Under the stability results of Theorem 3.4.3 for the adaptive controllers and identification models, it has been proved [15] that by using these three tools, only a finite number of switchings will take place during the control process and finally the plant will be controlled be a unique dominant controller.

The computational cost of this algorithm is another crucial factor for the success of control process. It is obvious that the computational burden depends on the number of T-S adaptive identification models. Due to the fact that in this strategy there is a uniform distribution of the identification models in a compact space Ξ, the larger this space is, the larger will be the number of the models. Theoretically, this number can be arbitrarily large, but in real problems there must be a compromise between the performance and the computational requirements of the controller. Thus, the number of models must be kept as small as possible by reducing the density of their distribution when the compact space Ξ is very large, providing at the same time a satisfactory performance. Simulations have shown that for simple systems the proposed method is faster than real time for a small number of models but it becomes gradually slower as the number of T-S identification models increases.

3.4.8 Another Approach with Hybrid T-S Multiple Models

In [25] the authors enhanced the TSMMASC architecture by using different types of multiple models and controllers together. More specifically, instead of using only adaptive estimation models, they proposed a scheme which contains $N - 2$ fixed T-S identification models $\left\{\mathcal{M}_f\right\}_{f=1}^{N-2}$, one free running adaptive model $\mathcal{M}_{ad}$ and one reinitialized adaptive model $\mathcal{M}_{adr}$ which are operating in parallel along with their corresponding fuzzy controllers. The main role of the fixed models is to provide the best possible approximation for the plant's parameters in the reinitialized adaptive model during the first steps of the control procedure. This strategy improves the systems' performance. A switching rule which is based on a cost criterion along with a hysteresis h and a $T_{\min}$ which is left to elapse between switchings, picks the best controller at every time instant. This way, the control signal to be applied is determined at every time instant by the model that approximates the plant best. At every step the parameters of both adaptive identification models and controllers are adapted simultaneously so that the asymptotic stability of the system is ensured. Similar to [24], in order to avoid singularity and infeasibility problems in the control signal, a projection technique and the NBCL strategy are used respectively.

The switching rule of this scheme is described as follows.

If $J_j(t) = \min_{k \in \Lambda}\{J_k(t)\}$, $\Lambda = \{1, \ldots, N\}$, and $J_j(t) + h \leq J_{cr}(t)$ is valid at least for the last evaluation of the cost criterion in the time interval $[t, t + T_{\min}]$ (depending on the algorithm step and the $T_{\min}$, there may be more than one evaluations in the time interval $[t, t + T_{\min}]$), then the controller $\mathcal{C}_j$ is chosen and tuned, provided that its tunable. That is, if $\mathcal{C}_j \{\mathcal{C}_{ad}, \mathcal{C}_{adr}\}$. Otherwise, the controller $\mathcal{C}_{cr}$ remains active, meaning that it is the ideal controller for the time instant $t + T_{\min}$. Here, $J_{cr}(t)$ is the index of the current active T-S model $\mathcal{M}_{cr}$. Note that $\mathcal{M}_j$, i.e. the model with the minimum cost criterion, may change during the evaluations in the time interval $[t, t + T_{\min}]$. The above procedure is repeated at every step. In case $\mathcal{M}_j \in \left\{\mathcal{M}_f\right\}_{f=1}^{N-2}$ i.e. $\mathcal{M}_j$ is a fixed model, the model $\mathcal{M}_{adr}$ reinitializes its parameter vector value, its cost criterion value and its state vector value according to the corresponding values of the dominant fixed model $\mathcal{M}_i$ and the entire adaptation process is continued.

The switching rule along with the adaptive laws that are given in [25], ensure the boundedness of all signals and the stability of the system. The analysis shows that one of the adaptive models ends up to be the dominant model in the proposed control architecture with the particular cost criterion (3.37). The cost criteria J_{ad} or J_{adr} of the adaptive models are bounded while the cost criteria of the fixed models grow in an unbounded way due to the fact that they contain the squared bounded errors along with their integral terms. Thus, there exists a finite time t_{ad} such that the system switches to one of the available adaptive controllers and stays there for all $t \geq t_{ad}$ [70]. The role of the reinitialized adaptive model is very significant because this model is designed to take the control role—if necessary—from a fixed model very quickly and lead the system to stability in case the free adaptive model does not have a satisfactory parameters initialization. Consequently, the main role of the fixed models is to provide the best possible approximation for the plant's parameters in the reinitialized adaptive model during the first steps of the control procedure.

The following example refers on the fuzzy control techniques that presented in this section. The control scheme is based on hybrid multiple models and the results confirm the aforementioned advantages of the TSMMASC architecture.

3.5 Numerical Example

Let the inverted pendulum system (3.50) which is a highly unstable system and is widely used as a benchmark control problem [39, 75]:

$$\dot{x}_1 = x_2$$
$$\dot{x}_2 = \frac{g \, \sin(x_1) - amlx_2^2 \sin(2x_1)/2 - a\cos(x_1)u}{4l/3 - aml\cos^2(x_1)} \qquad (3.50)$$

where x_1 denotes the angle (in radians) of the pendulum from the vertical, x_2 is the angular velocity, $g = 9.8$ m/s^2 is the gravity constant, m is the mass of the pendulum, M is the mass of the cart, $2\,l$ is the length of the pole, u is the force applied to the cart and $a = 1/(m + M)$. Two fuzzy rules are used to approximate the nonlinear system (3.50) when $x_1 \in (-\pi/2, \pi/2)$:

Rule 1: IF $x_1(t)$ is about 0 THEN $\dot{x}(t) = A_1x(t) + B_1u(t)$
Rule 2: IF $x_1(t)$ is about $\pm \pi/2$ THEN $\dot{x}(t) = A_2x(t) + B_2u(t)$,

where

$$A_1 = \begin{bmatrix} 0 & 1 \\ \frac{g}{4l/3-aml} & 0 \end{bmatrix}, \ B_1 - \begin{matrix} 0 \\ \frac{a}{4l/3-aml} \end{matrix}$$

$$A_2 = \begin{bmatrix} 0 & 1 \\ \frac{2g}{\pi(4l/3-aml\beta^2)} & 0 \end{bmatrix}, \ B_2 - \begin{matrix} 0 \\ \frac{a\beta}{4l/3-aml\beta^2} \end{matrix}$$

and $\beta = \cos(88°)$. The unknown parameters $\mathscr{E}_{AB}$ lie in the compact set Ξ:

$$\Xi = \{0.25 \le l \le 1.1, \ 0.5 \le m \le 4.5, \ 5.5 \le M \le 13\}$$

The control objective is to force the system (3.50) to follow the reference model (3.35) where:

$$Ad = \begin{bmatrix} 0 & 1 \\ -4 & -4 \end{bmatrix}$$

Using $Q_k = \mathbf{I}_{2\times2}$ in Lyapunov equation (3.45) one obtains:

$$P_k = \begin{bmatrix} 1.1250 & 0.1250 \\ 0.1250 & 0.1562 \end{bmatrix}, \ P_{ks}[0.1250\,0.1562]^{\mathrm{T}}.$$

Eight fixed fuzzy models $\{\mathscr{M}_k\}_{k=1}^{8}$, one free adaptive fuzzy model $\mathscr{M}_9$ and one reinitialized adaptive fuzzy model $\mathscr{M}_{10}$, along with their corresponding controllers are used in this simulation. The initial estimates $\hat{\mathbf{a}}^{ki}$, $\hat{b}^{ki}$ in the fixed models and the free adaptive model are distributed uniformly in the set Ξ. The values used in the adaptive law (3.47) are $r_1^{ki} = 2$ and $r_2^{ki} = 1$ for all $k \in \{ad, adr\}$, i. The values $a_c = 8$, $b_c = 1$ are used for the switching criterion (3.37), $T_{\min} = 0.01$ sec., $h = 0.01$ and the algorithm's step is $t_s = 0.01$ sec. The lower bounds for the b^i's are $b_0^1 = 0.015$ and $b_0^2 = 0.0015$. The initial states for the real plant, the free adaptive model, the fixed models and the reference model are $x = \hat{x}_k = [\pi/3\,0]^{\mathrm{T}}$ and $x_m = [\pi/4\,0]^{\mathrm{T}}$ and respectively.

The results of the proposed fuzzy control scheme are depicted in Fig. 3.4. In Fig. 3.4a–b, the states of the real plant (3.50) and the reference model (3.35) are given. The dashed line in all these figures depicts the reference model's state. In Fig. 3.4c and in the included subfigure, the control signal is depicted for the first five seconds and the first 0.5 seconds respectively. Finally, in Fig. 3.4d and in the included subfigure, the switching sequence of the controllers is depicted for the first five seconds and the first 0.3 seconds respectively. Due to the fact that the initial $J_k(t)$ is identical for all models $\mathscr{M}_k$, the same fixed controller $\mathscr{C}_4$ is chosen arbitrarily in order to provide the initial control signal to the system. It can be seen that the controlled system is stabilized and the states x_1, x_2 approximate the reference model's states after about five seconds. The control signal is smooth and there is no chattering effect. It can be noticed (Fig. 3.4d) that four controllers are

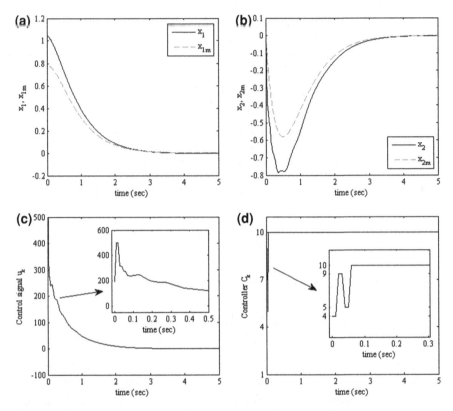

Fig. 3.4 **a–b** State responses, **c** control signal and **d** dominant controller sequence

used during the simulation. At the beginning controller $\mathscr{C}_4$ is used and remains active for two algorithm steps—due to the fact that $T_{\min}$ is used—although it is not proved to be the best controller. After that, the free adaptive controller $\mathscr{C}_9$ and the fixed controller $\mathscr{C}_5$ take the leadership sequentially for another two algorithm steps each, and finally the reinitialized adaptive controller $\mathscr{C}_{10}$ ends up to be the dominant controller of the system. It is obvious that the use of the hybrid multiple T-S models leads the system smoothly to the reference model's states. By using the fixed models instead of adaptive models we reduce the computational burden of the algorithm and make the controller simpler and quicker. In addition, the use of the free adaptive controller ensures the stability of the system and the reinitialized adaptive controller exploits the good initialization from the fixed models offering a faster convergence. The results are much the same for the T-S model of (3.50).

3.6 Conclusions

In this chapter, some of the most recent developments in the field of intelligent multiple models based adaptive switching control were presented along with some basic information about conventional adaptive and multiple models control methods. The advantages of using multiple models control schemes over the single models architectures were highlighted. More precisely, when the system to be controlled is highly uncertain, the single model adaptive control techniques may be insufficient. Using more than one identification models, one can improve the transient response and guarantee the stability of the system. Also the type and the synthesis of the multiple models is very important and depends on the system to be controlled. One can use fixed models only, adaptive models only or a combination of fixed and adaptive models. Generally, it is necessary to use at least one adaptive model in order to ensure the boundedness of all the signals and the stability of the system. The use of a reinitialized adaptive model along with free adaptive models offers many advantages like speed and accuracy. Intelligent multiple models control schemes offer extra advantages. Combining neural adaptive models and linear robust models, one can approximate and control any nonlinear system. Also, the use of T-S fuzzy adaptive models offers the possibility to use linear adaptive control techniques. The hybrid adaptive fuzzy structures are very efficient and faster with less computational cost. The convergence of all these algorithms is ensured by using the appropriate cost criteria and switching rules. Simulation results evince the efficiency of these control algorithms.

References

1. Athans, M., Castanon, D., Dunn, K., Greene, C., Lee, W., Sandell, J., Willsky, A.: The stochastic control of the F-8C aircraft using a multiple model adaptive control (MMAC) method–part I: equilibrium flight. Autom. Control IEEE Trans. **22**(5), 768–780 (1977)
2. Lainiotis, D.: Partitioning: a unifying framework for adaptive systems, I: estimation. Proc. IEEE **64**(8), 1126–1143 (1976)
3. Martensson, B.: Adaptive stabilization. Ph.D. Dissertation (1986)
4. Fu, M., Barmish, B.: Adaptive stabilization of linear systems via switching control. Autom. Control IEEE Trans. **31**(12), 1097–1103 (1986)
5. Miller, D., Davison, E.: An adaptive controller which provides Lyapunov stability. Autom. Control IEEE Trans. **34**(6), 599–609 (1989)
6. Middleton, R., Goodwin, G., Hill, D., Mayne, D.: Design issues in adaptive control. Autom. Control IEEE Trans. **33**(1), 50–58 (1988)
7. Morse, A., Mayne, D., Goodwin, G.: Applications of hysteresis switching in parameter adaptive control. Autom. Control IEEE Trans. **37**(9), 1343–1354 (1992)
8. Morse, A.: Supervisory control of families of linear set-point controllers. In: Proceedings of the 32nd IEEE Conference on Decision and Control, vol. 2, pp. 1055–1060 (1993)
9. Narendra, K., Balakrishnan, J.: Improving transient response of adaptive control systems using multiple models and switching. Autom. Control IEEE Trans. **39**(9), 1861–1866 (1994)

10. Narendra, K.S., Balakrishnan, J.: Adaptive control using multiple models. IEEE Trans. Autom. Control **42**, 171–187 (1997)
11. Hespanha, J., Liberzon, D., Morse, A.S., Anderson, B.D.O., Brinsmead, T.S., De Bruyne, F.: Multiple model adaptive control. part 2: switching. Int. J. Robust Nonlinear Control **11**(5), 479–496 (2001)
12. Kalkkuhl, J., Johansen, T., Ludemann, J.: Improved transient performance of nonlinear adaptive backstepping using estimator resetting based on multiple models. Autom. Control IEEE Trans. **47**(1), 136–140 (2002)
13. Freidovich, L.B., Khalil, H.K.: Logic-based switching for robust control of minimum-phase nonlinear systems. Syst. Control Lett. **54**(8), 713–727 (2005)
14. Cezayirli, A., Ciliz, M.K.: Indirect adaptive control of non-linear systems using multiple identification models and switching. Int. J. Control **81**(9), 1434–1450 (2008)
15. Ye, X.: Nonlinear adaptive control using multiple identification models. Syst. Control Lett. **57**(7), 578–584 (2008)
16. Kuipers, M., Ioannou, P.: Multiple model adaptive control with mixing. Autom. Control IEEE Trans. **55**(8), 1822–1836 (2010)
17. Baldi, S., Ioannou, P.A., Kosmatopoulos, E.B.: Adaptive mixing control with multiple estimators. Int. J. Adapt. Control Sig. Process. **26**, 800–820 (2012)
18. Han, Z., Narendra, K.: New concepts in adaptive control using multiple models. Autom. Control IEEE Trans. **57**(1), 78–89 (2012)
19. Narendra, K., Han, Z.: A new approach to adaptive control using multiple models. Int. J. Adapt. Control Sig. Process. **26**(8), 778–799 (2012)
20. Chen, L., Narendra, K.S.: Nonlinear adaptive control using neural networks and multiple models. Automatica **37**(8), 1245–1255 (2001)
21. Lee, C.Y., Lee, J.J.: Adaptive control for uncertain nonlinear systems based on multiple neural networks. Syst. Man Cybern. Part B: Cybern. IEEE Trans. **34**(1), 325–333 (2004)
22. Fu, Y., Chai, T.: Nonlinear multivariable adaptive control using multiple models and neural networks. Automatica **43**(6), 1101–1110 (2007)
23. Sofianos, N.A., Boutalis, Y.S., Christodoulou, M.A.: Feedback linearization adaptive fuzzy control for nonlinear systems: a multiple models approach. In: 19th Mediterranean Conference on Control Automation (MED), pp. 1453–1459. June 2011
24. Sofianos, N.A., Boutalis, Y.S.: Stable indirect adaptive switching control for fuzzy dynamical systems based on T-S multiple models. Int. J. Syst. Sci. **44**(8), 1546–1565, (2013)
25. Sofianos, N.A., Boutalis, Y.S., Mertzios, B.G., Kosmidou, O.I.: Adaptive switching control of fuzzy dynamical systems based on hybrid T-S multiple models. In: 6th IEEE International Conference on Intelligent Systems, Sofia, Bulgaria (2012)
26. Aseltine, J., Mancini, A., Sartune, C.: A survey of adaptive control systems. IRE Trans. Autom. Control **3**(6), 102–108 (1958)
27. Ioannou, P., Sun, J.: Robust Adaptive Control. Prentice Hall, Englewood Cliffs, New Jersey (1996)
28. Kalman, R.: Design of a self optimizing control system. Trans. ASME **80**, 468–478 (1958)
29. Osburn, P.V., Whitaker, H.P., Kezer, A.: New developments in the design of model reference adaptive control systems. Paper No. 61–39, Institute of the Aerospace Sciences (1961)
30. Parks, P.: Lyapunov redesign of model reference adaptive control systems. IEEE Trans. Autom. Control **11**, 362–367 (1966)
31. Egardt, B.: Stability of adaptive controllers. Lecture Notes in Control and Information Sciences, Springer, Berlin (1979)
32. Narendra, K., Lin, Y.H., Valavani, L.: Stable adaptive controller design, part II: proof of stability. IEEE Trans. Autom. Control **25**(3), 440–448 (1980)
33. Ioannou, P., Sun, J.: Theory and design of robust direct and indirect adaptive control schemes. Int. J. Control **47**(3), 775–813 (1988)
34. Wang, L.X.: Stable adaptive fuzzy control of nonlinear systems. Fuzzy Syst. IEEE Trans. **1**(2), 146–155 (1993)

35. Chen, B.S., Lee, C.H., Chang, Y.C.: H_∞ tracking design of uncertain nonlinear SISO systems: adaptive fuzzy approach. Fuzzy Syst. IEEE Trans. **4**(1), 32–43 (1996)
36. Andersen, H., Lotfi, A., Tsoi, A.: A new approach to adaptive fuzzy control: the controller output error method. Syst. Man Cybern. Part B: Cybern. IEEE Trans. **27**(4), 686–691 (1997)
37. Yang, Y., Ren, J.: Adaptive fuzzy robust tracking controller design via small gain approach and its application. Fuzzy Sys. IEEE Trans. **11**(6), 783–795 (2003)
38. Qi, R., Brdys, M.: Adaptive fuzzy modelling and control for discrete-time nonlinear uncertain systems. In: Proceedings of the American Control Conference 2005, vol. 2, pp. 1108–1113. June 2005
39. Park, C.W., Cho, Y.W.: T-S model based indirect adaptive fuzzy control using online parameter estimation. Syst. Man Cybern. Part B: Cybern. IEEE Trans. **34**(6), 2293–2302 (2004)
40. Park, C.W., Park, M.: Adaptive parameter estimator based on T-S fuzzy models and its applications to indirect adaptive fuzzy control design. Inf. Sci. **159**(1–2), 125–139 (2004)
41. Hyun, C.H., Park, C.W., Kim, S.: Takagi-Sugeno fuzzy model based indirect adaptive fuzzy observer and controller design. Inf. Sci. **180**(11), 2314–2327 (2010)
42. Huang, Y.S., Huang, Z.X., Zhou, D.Q., Chen, X.X., Zhu, Q.X., Yang, H.: Decentralised indirect adaptive output feedback fuzzy H_∞ tracking design for a class of large-scale nonlinear systems. Int. J. Syst. Sci. **43**(1), 180–191 (2012)
43. Chen, P.C., Wang, C.H., Lee, T.T.: Robust adaptive self-structuring fuzzy control design for nonaffine, nonlinear systems. Int. J. Syst. Sci. **42**(1), 149–169 (2011)
44. Huang, Y.S., Wu, M., He, Y., Yu, L.L., Zhu, Q.X.: Decentralized adaptive fuzzy control of large-scale nonaffine nonlinear systems by state and output feedback. Nonlinear Dyn. **69**(4), 1665–1677 (2012)
45. Fischle, K., Schroder, D.: An improved stable adaptive fuzzy control method. Fuzzy Sys. IEEE Trans. **7**(1), 27–40 (1999)
46. Liu, B.D., Chen, C.Y., Tsao, J.Y.: Design of adaptive fuzzy logic controller based on linguistic-hedge concepts and genetic algorithms. IEEE Trans. Syst. Man Cybern. Part B **31**(1), 32–53 (2001)
47. Ordonez, R., Zumberge, J., Spooner, J., Passino, K.: Adaptive fuzzy control: experiments and comparative analyses. Fuzzy Syst. IEEE Trans. **5**(2), 167–188 (1997)
48. Sanner, R., Slotine, J.J.: Gaussian networks for direct adaptive control. Neural Netw. IEEE Trans. **3**(6), 837–863 (1992)
49. Rovithakis, G., Christodoulou, M.: Adaptive control of unknown plants using dynamical neural networks. Syst. Man Cybern. IEEE Trans. **24**(3), 400–412 (1994)
50. Narendra, K.S., Mukhopadhyay, S.: Adaptive control of nonlinear multivariable systems using neural networks. Neural Netw. **7**(5), 737–752 (1994)
51. Polycarpou, M.: Stable adaptive neural control scheme for nonlinear systems. Autom. Control IEEE Trans. **41**(3), 447–451 (1996)
52. Polycarpou, M.M., Mears, M.J.: Stable adaptive tracking of uncertain systems using nonlinearly parametrized on-line approximators. Int. J. Control **70**(3), 363–384 (1998)
53. Kwan, C., Lewis, F.: Robust backstepping control of nonlinear systems using neural networks. Syst. Man Cybern. Part A: Syst. Hum. IEEE Trans. **30**(6), 753–766 (2000)
54. Ge, S., Wang, J.: Robust adaptive neural control for a class of perturbed strict feedback nonlinear systems. Neural Netw. IEEE Trans. **13**(6), 1409–1419 (2002)
55. Ge, S., Wang, C.: Adaptive neural control of uncertain MIMO nonlinear systems. Neural Netw. IEEE Trans. **15**(3), 674–692 (2004)
56. Kostarigka, A., Rovithakis, G.: Prescribed performance output feedback/observer-free robust adaptive control of uncertain systems using neural networks. Syst. Man Cybern. Part B: Cybern. IEEE Trans. **41**(6), 1483–1494 (2011)
57. Ho, D., Li, J., Niu, Y.: Adaptive neural control for a class of nonlinearly parametric time-delay systems. Neural Netw. IEEE Trans. **16**(3), 625–635 (2005)
58. Chen, W.S., Li, J.M.: Adaptive output-feedback regulation for nonlinear delayed systems using neural network. Int. J. Autom. Comput. **5**, 103–108 (2008)

59. Kar, I., Behera, L.: Direct adaptive neural control for affine nonlinear systems. Appl. Soft Comput. **9**(2), 756–764 (2009)
60. Zhang, T.P., Zhu, Q., Yang, Y.Q.: Adaptive neural control of non-affine pure-feedback nonlinear systems with input nonlinearity and perturbed uncertainties. Int. J. Syst. Sci. **43**(4), 691–706 (2012)
61. Da, F., Song, W.: Fuzzy neural networks for direct adaptive control. Ind. Electron. IEEE Trans. **50**(3), 507–513 (2003)
62. Liu, X.J., Lara-Rosano, F., Chan, C.: Model-reference adaptive control based on neurofuzzy networks. Syst. Man Cybern. Part C: Appl. Rev. IEEE Trans. **34**(3), 302–309 (2004)
63. Boutalis, Y., Theodoridis, D., Christodoulou, M.: A new neuro-FDS definition for indirect adaptive control of unknown nonlinear systems using a method of parameter hopping. Neural Netw. IEEE Trans. **20**(4), 609–625 (2009)
64. Theodoridis, D., Boutalis, Y., Christodoulou, M.: Dynamical recurrent neuro-fuzzy identification schemes employing switching parameter hopping. Int. J. Neural Syst. **22**(2), 16 (2012)
65. Theodoridis, D., Boutalis, Y., Christodoulou, M.: Direct adaptive neuro-fuzzy trajectory tracking of uncertain nonlinear systems. Int. J.Adapt. Control Sig. Process. **26**(7), 660–688, (2012)
66. Lee, C.H., Chung, B.R. Adaptive backstepping controller design for nonlinear uncertain systems using fuzzy neural systems. Int. J. Syst. Sci. **43**(10), 1855–1869 (2012)
67. Ioannou, P.A., Kokotovic, P.V.: Instability analysis and improvement of robustness of adaptive control. Automatica **20**(5), 583–594 (1984)
68. Liberzon, D.: Switching in Systems and Control. Birkhauser Boston (2003)
69. Narendra K., Han Z.: Location of models in multiple-model based adaptive control for improved performance. In: 2010 American Control Conference (ACC), pp. 117–122 (2010)
70. Narendra, K., George, K.: Adaptive control of simple nonlinear systems using multiple models. In: 2002 American Control Conference, vol. 3, pp. 1779–1784 (2002)
71. Ciliz, M.K., Cezayirli, A.: Increased transient performance for the adaptive control of feedback linearizable systems using multiple models. Int. J. Control **79**(10), 1205–1215 (2006)
72. Takagi, T., Sugeno, M.: Fuzzy identification of systems and its applications to modeling and control. Syst.Man Cybern. IEEE Trans. **15**, 116–132 (1985)
73. Ioannou, P, Fidan, B.: Adaptive Control Tutorial. SIAM, Philadelphia (2006)
74. Kang, H.J., Kwon, C., Lee, H., Park, M.: Robust stability analysis and design method for the fuzzy feedback linearization regulator. Fuzzy Syst. IEEE Trans. **6**(4), 464–472 (1998)
75. Wang, H., Tanaka, K., Griffin, M.: An approach to fuzzy control of nonlinear systems: stability and design issues. Fuzzy Syst. IEEE Trans. **4**(1), 14–23 (1996)

Chapter 4
A Computational Intelligence Approach to Software Component Repository Management

D. Vijay Rao and V. V. S. Sarma

Abstract Software productivity has been steadily increasing but not enough to close the gap between the demands placed on the software industry and what the state of the practice can deliver. With the increasing dependence of systems on software with increasing complexity and challenging quality requirements, software architectural design to support reuse has become an important development activity and this research domain is rapidly evolving. In the last decades, software architecture optimization methods, which aim to automate the search for an optimal architecture design with respect to a set of quality attributes, have proliferated in literature with the major paradigms of development being as *Design for Reuse* and *Design with Reuse*. Classification, storing, and selection of components from a reuse library are considered as a key success factors for software reuse projects, especially when reuse involves also software artifacts, besides code. Correct component classification helps address several problems, besides reuse, such as code comprehension for reverse engineering, dynamic domain modeling, evaluation of programming language dependencies, and usage patterns. In this chapter, we propose a computational intelligence approach using Rough-Fuzzy hybridization techniques to retrieve software components stored as a case-base. The integration of Case-based Reasoning and Decision theory based on Computational Intelligence techniques has shown its usefulness in the retrieval and selection of reusable software components from a software components repository. Software components are denoted by cases with a set of features, attributes, and relations of a given situation and its associated outcomes. These are taken as inputs to a Decision Support tool that classifies the components as *adaptable to the given situation with membership values for the decisions*. In this novel approach, CBR and DSS (based on Rough-Fuzzy sets) have been applied successfully to the software engineering domain to address the problem of retrieving suitable components for reuse from the case data repository. The use of rough-fuzzy sets

D. Vijay Rao (✉) · V. V. S. Sarma
Department of Computer Science and Automation, Indian Institute of Science,
Bangalore 560012, India
e-mail: doctor.rao.cs@gmail.com

V. E. Balas et al. (eds.), *Innovations in Intelligent Machines-5*,
Studies in Computational Intelligence 561, DOI: 10.1007/978-3-662-43370-6_4,
© Springer-Verlag Berlin Heidelberg 2014

increase the likelihood of finding the suitable components for reuse when exact matches are not available or are very few in number. This classification is based on Rough-Fuzzy set theory and the methodology is explained with illustrations.

4.1 Introduction

As software costs continue to represent an increasing share of computer system costs, and as software faults continue to be responsible for many expensive failures, nothing short of an order-of-magnitude improvement in software quality and development productivity will save the software industry from its perennial state of crisis. With significant industrial demand for software systems with increasing complexity and challenging quality requirements, software architecture design to support reuse has become an important development activity and the research domain is rapidly evolving [1, 20, 23, 28, 46, 15, 16, 18, 37]. In the last decades, reuse oriented development and software architecture optimization methods, which aim to automate the search for an optimal architectural design with respect to a set of quality attributes, have proliferated in literature with the major paradigms of development being as Design for Reuse and Design with Reuse [24–26, 46, 38, 28, 29, 31]. Classification, storing, and selection of components from a reuse library are considered as a key success factors for software reuse projects, especially when reuse involves also software artifacts, besides code [1, 20, 23]. Correct component classification helps address several problems, besides reuse, such as code comprehension for reverse engineering, dynamic domain modeling, evaluation of programming language dependencies, and usage patterns. In this chapter, we propose a computational intelligence approach using Rough-Fuzzy hybridization techniques to retrieve software components stored as a case-base.

Research about classification and retrieval techniques for software components has undergone a notable development in recent years. The original view of software reuse and components in object oriented technology were centered on reuse of single components. In this view, the critical issues are the identification of the appropriate granularity of the reusable component and the understanding and adaptation of reuse components to new application requirements. Meanwhile some organizations have already based their development process on large internally developed software libraries. Various projects and commercial systems exist, supporting the concept of large reusable component malls; a review of techniques for library management is presented in [27, 28, 29, 31, 38, 46]. Among the most promising approaches toward reuse, the paradigm of application frameworks is gaining wide acceptance in the object-oriented community [1–7, 17, 42–45]. Application frameworks are application skeletons that can be considered as components, in that they are sold as products and that an application can use various frameworks. Currently both styles of reuse, that is, traditional, library-based reuse and framework-based reuse, exist jointly in real development environments, and borders are often blurred. Whatever the reuse technique, the problem arises of locating and adapting the useful candidates on the basis of some specification of their behavior [11, 22, 30, 49]. These problems have recently assumed relevance in projects requiring access to large-scale systems of components and frameworks, as reported in Baumer et al. [4]. Damiani et al. [9] proposes two models for classification and retrieval of class libraries and application frameworks, stored in an object-oriented code base. These models are based on software descriptors describing the behavioral properties and non-behavioral characteristics of code. A novel approach where components are stored as cases in a case database of components and each component's reuse history in past applications is also stored in the case database. We use a case based reasoning approach that relies on past, similar cases (components) to find solutions to the current problem requirements. The selection of candidate components used in previous projects for reuse in the current application is modeled as Rough sets, Fuzzy sets, Rough-Fuzzy sets and Fuzzy-Rough sets which are useful soft computing tools with the distinguishing characteristic that they provide approximate solutions to approximately formulated problems. Existing approaches to software component retrieval cover a wide spectrum of component encoding methods and search or matching algorithms. These approaches strike different balances between complexity and cost on one hand and retrieval quality on the other, are classified by increasing order of complexity:

- Text-based "encoding" and retrieval,
- Lexical descriptor-based,
- Specification-based encoding and retrieval [29].

With text-based encoding and retrieval, the textual representation of a component is used as an implicit functional descriptor. Arbitrarily complex string search expressions supplied by the reuser are matched against the textual representation. With lexical descriptor-based encoding, each component is assigned a set of key phases that tell what the component is about. Specification-based encoding and retrieval comes closest to achieving full equivalence between what a component is and does, and how it is encoded. With text and lexical descriptor-based methods, retrieval algorithms treat queries and codes as mere symbols, and any meaning assigned to queries, component codes, or the extent of match between them is external to the encoding language. Further, being natural language based, the codes are inherently ambiguous and imprecise. By contrast, specific languages have their own semantics within which the fitness of a component to a query can be formally inter-preted. Mili et al. [30] describe a method for organizing and retrieving software components that uses relational specifications of program and refinement ordering between them. Damiani et al. [9] describe a hierarchy-aware classification schema for object-oriented code, where software components are classified according to their behavioral characteristics, such as provided services, employed algorithms, and needed data.

In most situations, the designer faces two major issues: how to deal with unknown problem features, and how to make decisions in the presence of these unknowns. Our methodology views retrieval and selecting components from a case-based repository as a decision problem. If the case (component) retrieved is very similar to the problem being solved (selecting a component needed for reuse), adaptation and testing will be easy. If the case is dissimilar, it will require a higher effort of adaptation [2–7, 14, 21, 24, 33, 34, 35, 38 39, 41, 42, 48]. Case-based Reasoning (CBR) systems can retrieve similar cases even in instances of uncertainty. To deal with uncertainties in CBR, we treat the selection of the best case (component that is reusable 'as-is') as a decision-making problem. The task becomes to analyse and incorporate uncertainties, and characteristics of the components and preferences so that the factors influencing a decision are part of the case retrieval process. Rough-fuzzy sets [12, 13] are used as a basis for decision-making in this context. In this chapter, a CBR system is proposed as a repository of reusable software components. Every reusable component with its design attributes and specifications is used as a case in a CBR system. The retrieval of components is based on their match in behavior/specification/reuse history. The component match in several applications is classified using rough-fuzzy sets, based on the similarity of the attributes. A decision based on rough-fuzzy sets is applied to the history of a component's use in different applications to determine its potential reuse in the current application.

4.2 A Computational Intelligence Based Approach Using Rough and Fuzzy Sets to Model the Component Retrieval Problem

In any classification task the aim is to form various classes where each class contains objects that are not noticeably different. These indiscernible or non-distinguishable objects can be viewed as basic building blocks (concepts) used to build up a knowledge base about the real world. For example, if the objects are classified according to color (red, black) and shape (triangle, square, circle), then the classes are (red triangle) (red square) (red circle) (black triangle) (black square) (black circle). Thus, these two attributes make a 'partition' in the set of objects and the universe becomes coarse. But if two red triangles of different sizes (small, big) belong to different classes, then it is not possible to correctly classify these two red triangles based on the two attributes of color and shape. This kind of uncertainty is referred to as rough uncertainty [8, 9, 10, 32, 33, 46]. The rough uncertainty is formulated in terms of rough sets. Fuzzy sets, a generalization of classical sets, is a mathematical tool to model the vagueness present in the human classification mechanism. In the classical set theory, an element of the universe either belongs to or does not belong to a set (i.e. crisp belongingness). In fuzzy sets, the belongingness of the element can be anything in between 'yes' or 'no' [19, 50, 51]. In fuzzy sets, each granule of knowledge can have only one membership value to a particular class. Rough sets assert that each granule may have different membership values to the same class. Fuzzy sets deal with overlapping classes and fine concepts; whereas rough sets deal with non-overlapping classes and coarse concepts. In a classification task, the indiscernibility relation partitions the input pattern set to form several equivalence classes. These equivalence classes try to approximate the given output class. When this approximation is not proper, the roughness is generated. However, the output classes (concepts) may have ill-defined boundaries. Thus, both roughness and fuzziness appear here due to the indiscernibility relation in the input pattern set (attributes) and the vagueness in the output class (decisions) respectively. To model such situations where both vagueness and approximation are present, rough-fuzzy sets are proposed [13].

4.3 Rough Sets

Let R be an equivalence relation on a universal set X. Moreover, let X/R denote the family of all equivalence classes induced on X by R. One such equivalence class in X/R, that contains $x \in X$, is designated by $[x]_R$. For any output class $A \subseteq X$, we can define the lower $\underline{R}(A)$ and upper $\overline{R}(A)$ approximations, which approach A as closely as possible from inside and outside, respectively. Here, $\underline{R}(A) = \cup\{[x]_R \mid [x]_R \subseteq A, x \in X\}$ is the union of all equivalence classes in X/R that are contained in A, and $\overline{R}(A) = \cup\{[x]_R \mid [x]_R \cap A \neq \phi, x \in X\}$ is the union of all equivalence classes in X/R that overlap with A. A rough set $R(A) = \langle \underline{R}(A), \overline{R}(A) \rangle$ is a representation of the given set A by $\underline{R}(A)$ and $\overline{R}(A)$. The set $BN(A) = \overline{R}(A) - \underline{R}(A)$ is a rough description of the boundary of A by the equivalence classes of X/R. The approximation is rough uncertainty free if $\underline{R}(A) = \overline{R}(A)$. Thus, when all the patterns (components) from an equivalence class do not carry the same output class label, rough ambiguity is generated as a manifestation of the one-to-many relationship between that equivalence class and the output class labels. The *rough membership function* $r_A(x) : A \rightarrow [0,1]$ of a pattern $x \in X$ for the output class A is defined by

$$r_A(x) = \frac{\| [x]_R \cap A \|}{\| [x]_R \|},$$

where, $\|A\|$ denotes the cardinality of the set A.

4.4 Fuzzy Sets

In traditional two-state classifiers, where a class A is defined as a subset of a universal set X, any input pattern $x \in X$ can either be a member or not be a member of the given class A. This property of whether or not a pattern x of the universal set belongs to the class A can be defined by a *characteristic function* $\mu_A : X \to \{0, 1\}$ as follows:

$$\mu_A(x) = \begin{cases} 1 \; \textit{iff } x \in A \\ 0 \; \textit{iff } x \notin A \end{cases}$$

In real life situations, however, boundaries between the classes may be overlapping. Hence, it is uncertain whether an input pattern belongs totally to the class A. To take care of such situations, in fuzzy sets, the concept of characteristic function has been modified to *fuzzy membership function* $\mu_A : X \to [0, 1]$. This function is called membership function.

4.5 Rough-Fuzzy Sets

Let X be a set and R be an equivalence relation defined on X and the output class $D_c \subseteq X$ be a fuzzy set. A rough-fuzzy set [13, 40] is a tuple

$$R(D_c) = \langle \underline{R}(D_c), \overline{R}(D_c) \rangle \tag{4.1}$$

where the lower approximation $\underline{R}(D_c)$ and the upper approximation $\overline{R}(D_c)$ of D_c are fuzzy sets of X/R, with their corresponding membership functions defined by

$$\mu_{\underline{R}(D_c)}([x]_R) = \inf\{\mu_{D_c}(x) | x \in [x]_R\}, \forall x \in X, \tag{4.2}$$

and

$$\mu_{\overline{R}(D_c)}([x]_R) = \sup\{\mu_{D_c}(x) | x \in [x]_R\}, \forall x \in X. \tag{4.3}$$

The rough-fuzzy membership function of a component $x \in X$ for the fuzzy output class $D_c \subseteq X$ is defined by

$$l_{D_c}(x) = \frac{|F \cap D_c|}{|F|}, \tag{4.4}$$

where $F = [x]_R$ and $|D_c|$ means the cardinality of the fuzzy set D_c and is defined as

$$|D_c| = \sum_{x \in X} \mu_{D_c}(x). \tag{4.5}$$

The intersection operation on A and B is defined as

$$\mu_{A \cap B}(x) = \min\{\mu_A(x), \mu_B(x)\}, \forall x \in X. \tag{4.6}$$

4.6 Fuzzy-Rough Sets

When the equivalence classes are not crisp, they are in form of fuzzy clusters $\{F_1, F_2, \ldots, F_H\}$ generated by a fuzzy weak partition of the input set X. The term fuzzy weak partition means that each F_j, $j \in \{1, 2, \ldots, H\}$ is a normal fuzzy set, i.e., $\sup_x \mu_{F_j}(x) = 1$ and $\inf_x \max_j \mu_{F_j}(x) > 0$ while $\sup_x \min \{\mu_{F_i}(x), \mu_{F_j}(x)\} < 1$; $\forall i, j \in \{1, 2, \ldots, H\}$.

Here $\mu_{F_j}(x)$ is the fuzzy membership function of the pattern x in the cluster F_j. In addition, the output classes C_c, $c = \{1, 2, \ldots, C\}$ may be fuzzy too. Given a weak fuzzy partition $\{F_1, F_2, \ldots, F_H\}$ on X, the description of any fuzzy set C_c by means of the fuzzy partitions under the form of an upper and a lower approximation $\underline{C_c}$ and $\overline{C_c}$ is as follows:

$$\mu_{\underline{C_c}}(F_j) = \inf_{x \in C_c} \max\{1 - \mu_{F_j}(x), \mu_{C_c}(x)\} \ \forall x$$

$$\mu_{\overline{C_c}}(F_j) = \sup_{x \in C_c} \min\{\mu_{F_j}(x), \mu_{C_c}(x)\} \ \forall x$$

The tuple $\langle \underline{C_c}, \overline{C_c} \rangle$ is called a fuzzy-rough set. Here, $\mu_{C_c}(x) = \{0, 1\}$ is the fuzzy membership of the input x to the class C_c. Fuzzy-roughness appears when a fuzzy cluster contains patterns that belong to different classes.

If the equivalent classes form the fuzzy clusters $\{F_1, F_2, \ldots, F_H\}$ the each fuzzy cluster can be considered as a fuzzy linguistic variable. The rough-fuzzy membership function can thus be generalized as

$$\tau_{C_c}^f(x) = \frac{1}{\sum_j \mu_{F_j}(x)} \sum_{j=1}^{H} \mu_{F_j}(x) \ell_{C_c}^j, \quad \text{if } \exists j \text{ with } \mu_{F_j}(x) > 0$$

$$= 0, \text{otherwise}$$

Here, $\ell_{C_c}^j = \frac{\|F_j \cap C_c\|}{\|F_j\|}$ and the term $\frac{1}{\sum_j \mu_{F_j}(x)}$ in the above definition normalizes the fuzzy membership $\mu_{F_j}(x)$, and therefore called constrained fuzzy-rough membership function.

4.7 A Case Study of the Design and Development of Air Warfare Simulation Systems

Design and development of large scale military simulations for training and operational analyses need to be adaptive to the frequent changes in doctrines, tactics, upgrades in weapon systems capabilities' and strategies in warfare. Hence, it becomes imperative to build flexibility and agility in the development product and processes. An engineering approach to such development is to use an architectural approach supporting reuse at different levels stages in the system development lifecycle by distinguishing between the commonality and variability aspects of development for later reuse in families of software product development.

Air Warfare Simulation Systems (AWSS) has been designed and developed with the purpose to simulate wide range of military air operations [23] such as offensive/defensive counter air missions, counter surface force operations, air defense missions, and combat support operations between two or more opposing forces.

A thorough analysis and evaluation of the M&S architecture was done before finalising the architectural framework [1] is shown in Fig. 4.1. A wide range of military operations can be planned using this system to meet the objectives like destruction of a synthetically generated target such as an airfield, vital bridge, refinery, power plant, army installations or other locations of strategic importance. Extensive set of game rules have been developed over the period of time to simulate these operations under different weather and terrain conditions to generate realistic results in a quantitative manner. This system is being used for training and operational analysis and also used as a decision support tool for different echelons of the military planners. AWSS provides a platform for deployment of resources, weapon target matching, weaponeering assessment, force planning, force execution, damage assessment, quantitative results analysis and displaying reasoning for generating outcomes which is crucial for debriefing and learning purpose. It also computes the attrition rates, statistics of various operations and in-depth history of various events generated during the simulation which helps in analysis and validation of tactics and various operational objectives (Fig. 4.1).

A case database containing the history of reusable components has been maintained at CASSA for several software projects on war-gaming and simulation built over the years. Of the 32 attributes in the information table (Table 4.1), we selected 8 attributes of the past applications of component C (Specifications, Adaptability for reuse, Domain, Quality, Functional usage, Criticality, Resource utilization, and Certified) (Table 4.2). Each of these features are searched, matched, and retrieved by the CBR (Figs. 4.1 and 4.2). The applications of a component C labeled C_1 to C_n that match the attributes of the current application are retrieved. The decision that was taken for reusing the component C in an application is shown in the "Reuse/ Adapt decision" column denoting the level of reuse. Consider a search query for component's application history that resulted in 10 matches, C_1 to C_{10}. This search with 10 likely candidate applications to choose from, are displayed by the CBR. The developer who wants to reuse the component in the present application has to decide on a past application(s) that best matches with the present application. These candidate applications (obtained by the CBR search) are stored in the information table with decisions on the level of reuse [(Table 4.3) and (Fig.4.3)] as:

a) The component is reused as is (**as-is**).
b) The component is adapted with minor code modification (**adapt-comp**).

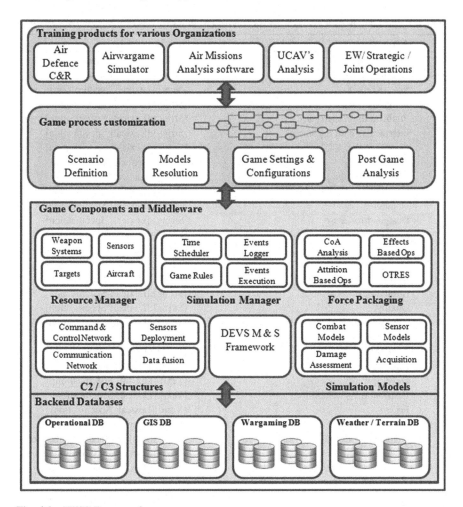

Fig. 4.1 AWSS Framework

c) The component is adapted with a modified specification with design modification (**adapt-spec**).
d) The component is developed afresh (**new**).

4.7.1 Use of Rough-Fuzzy Sets

In the information table, roughness occurs because of different decisions on the level of reuse taken for two similar applications; whereas fuzziness in the decision (on the level of reuse) occurs due to overlapping, ill-defined boundaries. Table 4.4

Table 4.1 Information of a software component stored in a case-base

#	Attribute	Search method
1	Component name	[String]:[Text; Lexical]
2	Alternative name	[String]:[Text; Lexical]
3	Size (KLOC)	[Num]:[Lexical; Specification]
4	Language used	[String]:[Lexical; Specification]
5	Platform developed on	[String]:[Lexical; Specification]
6	Interface(s) with arguments	[String]:[Lexical; Specification]
7	Sample usage	[String]:[Lexical; Specification]
8	Complexity	[Num]:[Lexical; Specification]
9	Functional usage profile	[String]:[Lexical; Specification]
10	Criticality in application	[String]:[Lexical; Specification]
11	Domain	[String]:[Lexical; Specification]
12	Designed for reuse	[String]:[Lexical; Specification]
13	Specifications	[String]:[Lexical; Specification]
14	Certification	[String]:[Lexical; Specification]
15	Test plan/report	[Text]:[Lexical; Specification]
16	Known bugs	[Text]:[Lexical; Specification]
17	Limitations	[String]:[Lexical; Specification]
18	Quality	[String]:[Lexical; Specification]
19	<Pre> {Specification} <Post> conditions	[String]:[Lexical; Specification]
20	History of component's reuse	[Text]:[Lexical; Specification]
21	Remarks	[Text]:[Lexical; Specification]
22	Effort (pm)	[Num]:[Lexical; Specification]
23	Developed by	[Text]:[Lexical; Specification]
24	Distribution/Availability	[Text]:[Lexical; Specification]
25	Price	[Num]:[Lexical; Specification]
26	Performance	[String]:[Lexical; Specification]
27	Algorithmic Complexity	[Num,String]:[Lexical; specification]
28	Resource utilization	[String]:[Lexical; Specification]
29	Test Coverage Statement coverage Branch coverage Definition-use	[String]:[Lexical; Specification]
30	Inheritance/Dynamic usage	[String]:[Lexical; Specification]
31	Data structures: Static: -and bounds/limits Dynamic: -and bounds/limits	[Num;String]:[Lexical; Specification]
32	Flexibility and Interoperability	[Text]:[Lexical; Specification]

reflects the difference of opinion among software developers regarding the level of reuse of component C in 10 applications. With a reuse history of a component in 10 past applications in the case-base, using rough-fuzzy sets, we compute the lower and upper membership values of similar applications. Equivalent classes for 10 applications that have same attributes are

$$\{\{C_1\}, \{C_2, C_6\}, \{C_3, C_4\}, \{C_5\}, \{C_7\}, \{C_8, C_9, C_{10}\}\}.$$

For decision D_1 (**as-is** reuse) the membership values of the equivalence classes are calculated using Eqs. (4.2) and (4.3) and Table 4.4.

Table 4.2 Attributes considered for Rough-Fuzzy sets and their descriptions

Attribute	Linguistic (match) variable	Description
1. Specification	Low	More than 70 % of functional requirements met
	Medium	30–70 % of functional requirements are met
	High	Less than 30 % functional requirements are met
2. Quality	Low	Defect density >10 defects/KLOC
	Medium	Defect density 5–10 defects/KLOC
	High	Defect density <5 defects/KLOC
3. Functional usage	Low	Called <15 times
	Medium	Called 15–45 times
	High	Called >45 times
4. Criticality	Low	Negligible effect on system functioning
	Medium	Degraded performance (settings/probability values)
	High	Halts the system from working, catastrophic failures
5. Domain	PL	Product line
	BC	Business/Commercial
	AF	Allied field
	MD	Misc. domain
6. Adaptable for reuse	Low	>10 person-weeks to adapt
	Medium	4–10 person-weeks to adapt
	High	<4 person-weeks to adapt
7. Resource utilization	Low	
	Medium	
	High	
8. Certified	Yes	
	No	
Reuse Decision:		
	As-is (D_1)	Reused with no additional effort to adapt
	Adapt-Comp (D_2)	Reused with change in code
	Adapt-Specs (D_3)	Reused by changing the requirement specification
	New (D_4)	Develop afresh

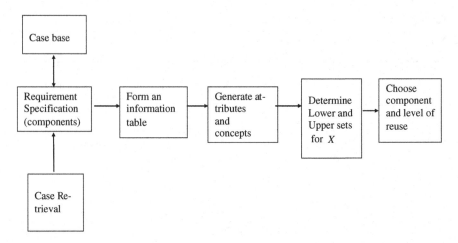

Fig. 4.2 Case (component) retrieval as a decision process based on Rough and Fuzzy Sets

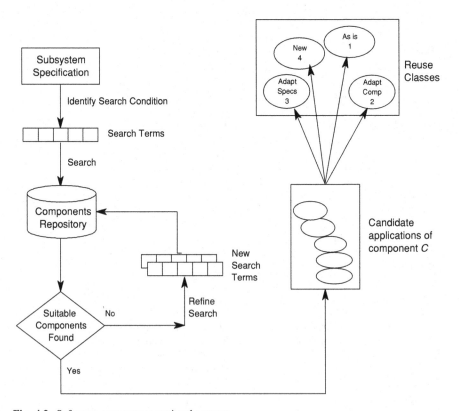

Fig. 4.3 Software components retrieval process

Table 4.3 Information table for 10 applications (from CBR search results)

Appln.	Specification match	Adapt for reuse	Domain	Quality	Functional usage	Criticality	Resource utilization	Certified	Reuse/Adapt in application
C_1	High	High	PL	High	High	High	Low	Yes	As-is
C_2	High	High	AF	High	High	Medium	Low	Yes	Adapt-Comp
C_3	Medium	High	PL	Medium	High	Medium	Low	Yes	As-is
C_4	Medium	High	PL	Medium	High	Medium	Low	Yes	Adapt-Comp
C_5	Low	Low	BC	Medium	Medium	Medium	Low	Yes	New
C_6	High	High	AF	High	High	Medium	Low	Yes	As-is
C_7	Low	High	MD	High	High	Medium	High	Yes	Adapt-Specs
C_8	High	Low	AF	High	High	Medium	Low	Yes	Adapt-Comp
C_9	High	Low	AF	High	High	Medium	Low	Yes	Adapt-Comp
C_{10}	High	Low	AF	High	High	Medium	Low	Yes	Adapt-Comp

Table 4.4 Table reflecting difference of opinion among software developers regarding reuse

Appln.	As-is (D_1)	Adapt-Comp (D_2)	Adapt-Specs (D_3)	New (D_4)	Actual decision taken
C_1	0.9	0.3	0.2	0.1	D1
C_2	0.3	0.9	0.3	0.1	D2
C_3	0.9	0.4	0.3	0.1	D1
C_4	0.6	0.6	0.3	0.1	D2
C_5	0.1	0.2	0.3	0.9	D4
C_6	0.8	0.3	0.2	0.1	D1
C_7	0	0.2	0.9	0.3	D3
C_8	0.3	0.9	0.3	0.1	D2
C_9	0.3	0.9	0.3	0.1	D2
C_{10}	0.3	0.9	0.3	0.1	D2

$$\mu_{\underline{R}(D_1)}([x]_R) = \inf\{\mu_{(D_1)}(\{C_1\})\} = 0.9$$
$$= \inf\{\mu_{(D_1)}(\{C_2, C_6\})\} = \inf(0.3, 0.8) = 0.3$$
$$= \inf\{\mu_{(D_1)}(\{C_3, C_4\})\} = \inf(0.9, 0.6) = 0.6$$
$$= \inf\{\mu_{(D_1)}(\{C_5\})\} = 0.1$$
$$= \inf\{\mu_{(D_1)}(\{C_7\})\} = 0$$
$$= \inf\{\mu_{(D_1)}(\{C_8, C_9, C_{10}\})\} = \inf(0.3, 0.3, 0.3) = 0.3$$

Therefore,

$$\mu_{\underline{R}(D_1)}([x]_R) = (0.9, 0.3, 0.6, 0.1, 0, 0.3).$$

$$\mu_{\underline{R}(D_1)}([x]_R) = \sup[\mu_{(D_1)}(C_1)] = 0.9$$
$$= \sup\{\mu_{(D_1)}(\{C_2, C_6\})\} = \sup(0.3, 0.8) = 0.8$$
$$= \sup\{\mu_{(D_1)}(\{C_3, C_4\})\} = \sup(0.9, 0.6) = 0.9$$
$$= \sup\{\mu_{(D_1)}(\{C_5\})\} = 0.1$$
$$= \sup\{\mu_{(D_1)}(\{C_7\})\} = 0$$
$$= \sup\{\mu_{(D_1)}(\{C_8, C_9, C_{10}\})\} = \sup(0.3, 0.3, 0.3) = 0.3$$

Therefore, $\mu_{\overline{R}(D_1)}([x]_R) = (0.9, 0.8, 0.6, 0.1, 0, 0.3)$.
Similarly, we calculate for decisions D_2, D_3, and D_4 to get

$$\mu_{\underline{R}(D_1)}([x]_R) = (0.9, 0.3, 0.6, 0.1, 0, 0.3)$$
$$\mu_{\overline{R}(D_1)}([x]_R) = (0.9, 0.8, 0.9, 0.1, 0, 0.3);$$
$$\mu_{\underline{R}(D_2)}([x]_R) = (0.3, 0.3, 0.4, 0.2, 0.2, 0.9)$$
$$\mu_{\overline{R}(D_2)}([x]_R) = (0.3, 0.9, 0.6, 0.2, 0.2, 0.9);$$
$$\mu_{\underline{R}(D_3)}([x]_R) = (0.2, 0.2, 0.3, 0.3, 0.9, 0.3)$$
$$\mu_{\overline{R}(D_3)}([x]_R) = (0.2, 0.3, 0.3, 0.3, 0.9, 0.3);$$
$$\mu_{\underline{R}(D_4)}([x]_R) = (0.1, 0.1, 0.1, 0.9, 0.3, 0.1)$$
$$\mu_{\overline{R}(D_4)}([x]_R) = (0.1, 0.1, 0.1, 0.9, 0.3, 0.1).$$

We need to determine those components that match the application's attributes and hence compute the lower and upper approximation of the rough-fuzzy set (information table) for the decisions denoting the level of reuse. The elements of the decisions (concepts) are

$$D_1 = \{C_1, C_3, C_6\}$$
$$D_2 = \{C_2, C_4, C_8, C_9, C_{10}\}$$
$$D_3 = \{C_7\}$$
$$D_4 = \{C_5\}$$

This is needed to determine the level of reuse of the component in a new application. Consider four new applications N1, N2, N3, N4 with the 8 attributes enumerated as

N1:   High, High, AF, High, High, Medium, Low, Yes
N2:   High, Low, MD, Low, Low, High, High, Yes
N3:   Medium, High, PL, Medium, High, Medium, Low, Yes
N4:   High, Low, AF, High, High, Medium, Low, Yes

Table 4.5 Rough-fuzzy membership values for decisions (10 applications)

#	Equivalence classes	l_{D_1}	l_{D_2}	l_{D_3}	l_{D_4}
1.	$\{C_1\}$	0.9	0.3	0.2	0.1
2.	$\{C_2,C_6\}$	0.55	0.6	0.25	0.1
3.	$\{C_3,C_4\}$	0.75	0.5	0.3	0.1
4.	$\{C_5\}$	0.1	0.2	0.3	0.9
5.	$\{C_7\}$	0	0.2	0.9	0.3
6.	$\{C_8,C_9,C_{10}\}$	0.3	0.9	0.3	0.1
N1	$\{C_2,C_6\}$	0.55	0.6	0.25	0.1
N2	–	–	–	–	–
N3	$\{C_3,C_4\}$	0.75	0.5	0.3	0.1
N4	$\{C_8,C_9,C_{10}\}$	0.3	0.9	0.3	0.1

The rough-fuzzy membership function (Eq. 4.4) is used to compute the membership value of a component's applications for decisions D_1, D_2, D_3, D_4.

$$C_{D_1} = \{(C_1,\ 0.9),\ (C_2,\ 0.3),\ (C_3,\ 0.9),\ (C_4,\ 0.6),\ (C_5,\ 0.1),\ (C_6,\ 0.8),$$
$$(C_7,\ 0),\ (C_8,\ 0.3),\ (C_9,\ 0.3),\ (C_{10},\ 0.3)\}.$$

$$l_{D_c}(C_1) = \frac{|\{C_1\} \cap C_{D_1}|}{|F|}$$

$$= \frac{\sum_{x \in X} \min\{\mu_{\{C_1\}}(x), \mu_{C_{D_1}}(x)\}}{1}$$

$$= \min\{\mu_{\{C_1\}}(C_1), \mu_{C_{D_1}}(C_1)\} + \cdots + \min\{\mu_{\{C_1\}}(C_{10}), \mu_{C_{D_1}}(C_{10})\}$$
$$= \min\{1,\ 0.9\} + 0 = 0.9$$

In a similar way, the other values are calculated. These are summarized in Table 4.5. The attributes of the new application N1 are matched with the past applications. The attributes of N1 match exactly with the equivalent class $\{C_2, C_6\}$. For this new application, we obtain the membership values for the decisions D_1, D_2, D_3, D_4 to determine the level of reuse of the component C's application $\{C_2, C_6\}$ from the Table 4.5.

This indicates that the new application N1's requirements are similar to applications C_2 and C_6 and they can be reused **as-is** (decision D_1) with membership value 0.55; and **adapt-comp** (decision D_2) with membership value 0.6 **adapt-specs** (decision D_3) with membership value 0.25; and **new** (decision D_4) with membership value 0.1. Application N2 does not match with any of the existing applications and hence it is decided to develop **new** (decision D_4). Similarly for applications N3 and N4 the values are calculated. The results are summarized in Table 4.5.

An important decision that arises at this point is whether the components arising from D_2, D_3, and D_4 decisions are introduced into the components case database, and if they would replace/coexist with the existing (versions of) components. This work is being worked in a separate forthcoming paper. In this chapter, for simplicity, we assume that they are all added to coexist with the existing components for future reuse. As the number of applications of component C increase, the level of reuse of components in new applications also changes. Consider the case when 20 applications of the same component C are available in the case-base. The information table with 20 applications is shown in Table 4.6. The membership values for decisions is shown in Table 4.7. Equivalence classes for the 20 applications that have the same attributes are

$$\{\{C_1\}, \{C_2, C_6\}, \{C_3, C_4, C_{19}\}, \{C_5, C_{11}\}, \{C_7, C_{17}\}, \{C_8, C_9, C_{10}, C_{20}\},$$
$$\{C_{12}, C_{18}\}, \{C_{13}\}, \{C_{14}, C_{16}\}, \{C_{15}\}\}$$

The decision sets are

$$D_1 = \{C_1,\ C_3,\ C_6, C_{16},\ C_{18},\ C_{19}, C_{20}\}$$
$$D_2 = \{C_2,\ C_4,\ C_8, C_9,\ C_{10}, C_{17}\}$$
$$D_3 = \{C_7, C_{11}\}$$
$$D_4 = \{C_5, C_{12},\ C_{13},\ C_{14}, C_{15}\}$$

Proceeding similarly as above the lower and upper membership values are calculated summarized.

$$\mu_{\underline{R}(D_1)}([x]_R) = (0.9,\ 0.3,\ 0.6,\ 0.1,\ 0,\ 0.3,\ 0,\ 0,\ 0,\ 0)$$
$$\mu_{\overline{R}(D_1)}([x]_R) = (0.9,\ 0.8,\ 0.9,\ 0.1,\ 0.3,\ 0.9,\ 0.9,\ 0,\ 0.9,\ 0);$$
$$\mu_{\underline{R}(D_2)}([x]_R) = (0.3,\ 0.3,\ 0.4,\ 0.2,\ 0.2,\ 0.4,\ 0.1,\ 0.1,\ 0.2,\ 0.1)$$
$$\mu_{\overline{R}(D_2)}([x]_R) = (0.3,\ 0.9,\ 0.6,\ 0.2,\ 0.9,\ 0.9,\ 0.6,\ 0.1,\ 0.3,\ 0.1);$$
$$\mu_{\underline{R}(D_3)}([x]_R) = (0.2,\ 0.2,\ 0.3,\ 0.3,\ 0.3,\ 0.2,\ 0.2,\ 0.4,\ 0.2,\ 0.2)$$
$$\mu_{\overline{R}(D_3)}([x]_R) = (0.2,\ 0.3,\ 0.3,\ 0.9,\ 0.9,\ 0.3,\ 0.3,\ 0.4,\ 0.4,\ 0.2);$$
$$\mu_{\underline{R}(D_4)}([x]_R) = (0.1,\ 0.1,\ 0,\ 0.3,\ 0.1,\ 0,\ 0,\ 0.9,\ 0,\ 0.9)$$
$$\mu_{\overline{R}(D_4)}([x]_R) = (0.1,\ 0.1,\ 0.1,\ 0.9,\ 0.3,\ 0.1,\ 0.9,\ 0.9,\ 0.9,\ 0.9).$$

The decision values for the new applications N1, N2, N3, N4 are summarized in Table 4.8. For N1 applications requirements, the attributes are same as those in equivalence class $\{C_2, C_6\}$ and hence on calculating the rough-fuzzy membership function for this class, the support for decision D1 and D2 are 0.55 and 0.6 respectively. These values do not change even after the case database is increased to 20 components. N2 application's attributes do not match any of the equivalence classes of 10 components database and hence the existing case base does not give any guidance on selecting the component. It is decided to develop it afresh (**new**)

Table 4.6 Information table for 20 applications (from CBR search results)

Appln.	Specifications match	Designed for reuse	Domain	Quality	Functional usage	Criticality	Resource utilization	Certified	Reuse/Adapt in application
C_1	High	High	PL	High	High	High	Low	Yes	As-is
C_2	High	High	AF	High	High	Medium	Low	Yes	Adapt-Comp
C_3	Medium	High	PL	Medium	High	Medium	Low	Yes	As-is
C_4	Medium	High	PL	Medium	High	Medium	Low	Yes	Adapt-Comp
C_5	Low	Low	BC	Medium	Medium	Medium	Low	Yes	New
C_6	High	High	AF	High	High	Medium	Low	Yes	As-is
C_7	Low	High	MD	High	High	Medium	High	Yes	Adapt-Specs
C_8	High	Low	AF	High	High	Medium	Low	Yes	Adapt-Comp
C_9	High	Low	AF	High	High	Medium	Low	Yes	Adapt-Comp
C_{10}	High	Low	AF	High	High	Medium	Low	Yes	Adapt-Comp
C_{11}	Low	Low	BC	Medium	Medium	Medium	Low	Yes	Adapt-Specs
C_{12}	High	No	MD	Low	Low	High	High	Yes	New
C_{13}	Low	Low	BC	High	High	High	High	No	New
C_{14}	Medium	Low	PL	Low	Low	Low	High	No	New
C_{15}	Medium	Low	MD	Low	Medium	Medium	High	No	New
C_{16}	Medium	Low	PL	Low	Low	Low	High	No	As-is
C_{17}	Low	High	MD	High	High	Medium	High	Yes	Adapt-Comp
C_{18}	High	No	MD	Low	Low	High	High	Yes	As-is
C_{19}	Medium	High	PL	Medium	High	Medium	Low	Yes	As-is
C_{20}	High	Low	AF	High	High	Medium	Low	Yes	As-is

Table 4.7 Table reflecting difference of opinion among software developers regarding reuse

Appln.	As-is (D1)	Adapt-Comp (D2)	Adapt-Specs (D3)	New (D4)	Actual decision taken
C_1	0.9	0.3	0.2	0.1	D1
C_2	0.3	0.9	0.3	0.1	D2
C_3	0.9	0.4	0.3	0.1	D1
C_4	0.6	0.6	0.3	0.1	D2
C_5	0.1	0.2	0.3	0.9	D4
C_6	0.8	0.3	0.2	0.1	D1
C_7	0	0.2	0.9	0.3	D3
C_8	0.3	0.9	0.3	0.1	D2
C_9	0.3	0.9	0.3	0.1	D2
C_{10}	0.3	0.9	0.3	0.1	D2
C_{11}	0.1	0.2	0.9	0.3	D3
C_{12}	0	0.1	0.2	0.9	D4
C_{13}	0	0.1	0.4	0.9	D4
C_{14}	0	0.2	0.4	0.9	D4
C_{15}	0	0.1	0.2	0.9	D4
C_{16}	0.9	0.3	0.2	0	D1
C_{17}	0.3	0.9	0.3	0.1	D2
C_{18}	0.9	0.6	0.3	0	D1
C_{19}	0.9	0.6	0.3	0	D1
C_{20}	0.9	0.4	0.2	0	D1

Table 4.8 Rough-fuzzy membership values for decisions (20 applications)

#	Equivalence classes	l_{D_1}	l_{D_2}	l_{D_3}	l_{D_4}
1.	$\{C_1\}$	0.9	0.3	0.2	0.1
2.	$\{C_2,C_6\}$	0.55	0.6	0.25	0.1
3.	$\{C_3,C_4,C_{19}\}$	0.8	0.54	0.3	0.066
4.	$\{C_5,C_{11}\}$	0.1	0.2	0.6	0.6
5.	$\{C_7,C_{17}\}$	0.15	0.55	0.6	0.2
6.	$\{C_8,C_9,C_{10},C_{20}\}$	0.45	0.78	0.28	0.075
7.	$\{C_{12},C_{18}\}$	0.45	0.35	0.25	0.45
8.	$\{C_{13}\}$	0	0.1	0.4	0.9
9.	$\{C_{14},C_{16}\}$	0.45	0.25	0.3	0.45
N1	$\{C_2,C_6\}$	0.55	0.6	0.25	0.1
N2	–	–	–	–	–
N3	$\{C_3,C_4,C_{19}\}$	0.8	0.54	0.3	0.066
N4	$\{C_8,C_9,C_{10},C_{20}\}$	0.45	0.78	0.28	0.075

and added to the existing case database. Application N3 changes its membership values for the decisions and the support for **as-is** reuse increases from 0.75 to 0.8; **adapt-comp** (decision D_2) with membership value from 0.5 to 0.54; **adapt-specs** (decision D_3) with membership value 0.3 (no change); **new** (decision D_4) with

Table 4.9 Fuzzy-Rough membership values for decisions (10 applications)

#	Fuzzy cluster	l_{D_1}	l_{D_2}	l_{D_3}	l_{D_4}
F1.	$\{C_1\}$	0.9	0.3	0.2	0.1
F2.	$\{C_2, C_6\}$	0.55	0.6	0.25	0.1
F3.	$\{C_3, C_4\}$	0.75	0.5	0.3	0.1
F4.	$\{C_5\}$	0.1	0.2	0.3	0.9
F5.	$\{C_7\}$	0	0.2	0.9	0.3
F6.	$\{C_8, C_9, C_{10}\}$	0.3	0.9	0.3	0.1
N1	$\{C_2, C_6\}$	0.55	0.6	0.25	0.1
N2	–	–	–	–	–
N3	$\{C_3, C_4\}$	0.75	0.5	0.3	0.1
N4	$\{C_8, C_9, C_{10}\}$	0.3	0.9	0.3	0.1

membership value decreasing from 0.1 to 0.066. Application N4 changes its membership values for the decisions and the support for **as-is** reuse increases from 0.3 to 0.45; **adapt-comp** (decision D_2) with membership value from 0.9 to 0.78; **adapt-specs** (decision D_3) with membership value 0.3 to 0.28; **new** (decision D_4) with membership value decreasing from 0.1 to 0.075. For new applications, as the case-database increases, the level of reuse improves (from **new** to **as-is**). As the case-database size increases, there is more **as-is** reuses than **new** development. This enables in a higher cost saving for the new application that reuses components from the case repository. The quantitative cost-benefit analyses for different levels of reuse of components from the components' repository are being worked out as a separate paper. A software tool called RuF Tool is developed as a decision support tool for reuse on Windows platform with MS-ACCESS as the back-end and Visual BASIC as the front-end. The components of past projects have been stored in this database as cases and RuFTool is used on this case database to retrieve the candidate components for reuse with membership values to support decision in reusing a component (Tables 4.9, 4.10).

4.7.2 Use of Fuzzy-Rough Sets

The case database in the Rough-Fuzzy set case study is used as the basis and membership functions for the components are obtained by the applicability of the component in the current requirements. This data is collected from the designers and quantified. The fuzzy clusters F_j of reusable component's that have for 10 components are:

$$F_j = \{C_1, C_2, C_3, C_4, C_5, C_6, C_7, C_8, C_9, C_{10}\}$$

Table 4.10 Rough-fuzzy membership values for decisions (20 applications)

#	Equivalence classes	l_{D_1}	l_{D_2}	l_{D_3}	l_{D_4}
F1.	$\{C_1\}$	0.9	0.3	0.2	0.1
F2.	$\{C_2, C_6\}$	0.55	0.6	0.25	0.1
F3.	$\{C_3, C_4, C_{19}\}$	0.8	0.54	0.3	0.066
F4.	$\{C_5, C_{11}\}$	0.1	0.2	0.6	0.6
F5.	$\{C_7, C_{17}\}$	0.15	0.55	0.6	0.2
F6.	$\{C_8, C_9, C_{10}, C_{20}\}$	0.45	0.78	0.28	0.075
F7.	$\{C_{12}, C_{18}\}$	0.45	0.35	0.25	0.45
F8.	$\{C_{13}\}$	0	0.1	0.4	0.9
F9.	$\{C_{14}, C_{16}\}$	0.45	0.25	0.3	0.45
N1	$\{C_2, C_6\}$	0.55	0.6	0.25	0.1
N2	–	–	–	–	–
N3	$\{C_3, C_4, C_{19}\}$	0.8	0.54	0.3	0.066
N4	$\{C_8, C_9, C_{10}, C_{20}\}$	0.45	0.78	0.28	0.075

$$\mu_{F_1} = (1, 0.2, 0.1, 0.2, 0.2, 0.1, 0.3, 0.1, 0.1, 0.2)$$
$$\mu_{F_2} = (0.2, 1, 0.1, 0.1, 0.2, 1, 0.2, 0.3, 0.2, 0.1)$$
$$\mu_{F_3} = (0.2, 0.1, 1, 1, 0.2, 0.3, 0.2, 0.1, 0.2, 0.1)$$
$$\mu_{F_4} = (0.2, 0.3, 0.2, 0.1, 1, 0.2, 0.3, 0.4, 0.1, 0.2)$$
$$\mu_{F_5} = (0.1, 0.2, 0.1, 0.4, 0.3, 0.1, 1, 0.2, 0.2, 0.1)$$
$$\mu_{F_6} = (0.2, 0.3, 0.1, 0.2, 0.1, 0.3, 0.1, 1, 1, 1).$$

The output class (decision) $D_1 \ldots D_4$ has its membership values as:

$$D = \{C_1, C_2, C_3, C_4, C_5, C_6, C_7, C_8, C_9, C_{10}\}$$

$$\mu_{D_1} = (0.9, 0.3, 0.9, 0.6, 0.1, 0.8, 0, 0.3, 0.3, 0.3)$$
$$\mu_{D_2} = (0.3, 0.9, 0.4, 0.6, 0.2, 0.3, 0.2, 0.9, 0.9, 0.9)$$
$$\mu_{D_3} = (0.2, 0.3, 0.3, 0.3, 0.3, 0.2, 0.9, 0.3, 0.3, 0.3)$$
$$\mu_{D_4} = (0.1, 0.1, 0.1, 0.1, 0.9, 0.1, 0.3, 0.1, 0.1, 0.1).$$

The fuzzy-Rough set is calculated using equations in Sect. 4.6 and the above membership values to obtain

1. **Decision D_1 (as-is reuse)**

$$D_1 = \{F1, F2, F3, F4, F5, F6\}$$

$$\mu_{\underline{D_1}} = (0.7, 0.3, 0.6, 0.1, 0, 0.3)$$
$$\mu_{\overline{D_1}} = (0.9, 0.8, 0.9, 0.3, 0.4, 0.3)$$

2. **Decision D_2 (adapt-comp reuse)**

$$D_2 = \{F1,\ F2,\ F3,\ F4,\ F5,\ F6\}$$

$$\mu_{\underline{D_2}} = (0.3,\ 0.3,\ 0.4,\ 0.2,\ 0.2,\ 0.7)$$

$$\mu_{\overline{D_2}} = (0.3,\ 0.9,\ 0.6,\ 0.4,\ 0.1,\ 0.9)$$

3. **Decision D_3 (adapt-specs reuse)**

$$D_3 = \{F1,\ F2,\ F3,\ F4,\ F5,\ F6\}$$

$$\mu_{\underline{D_3}} = (0.2,\ 0.2,\ 0.3,\ 0.3,\ 0.6,\ 0.3)$$

$$\mu_{\overline{D_3}} = (0.3,\ 0.3,\ 0.3,\ 0.3,\ 0.9,\ 0.3)$$

4. **Decision D_4 (new)**

$$D_4 = \{F1,\ F2,\ F3,\ F4,\ F5,\ F6\}$$

$$\mu_{\underline{D_4}} = (0.1,\ 0.1,\ 0.1,\ 0.6,\ 0.3,\ 0.1)$$

$$\mu_{\overline{D_4}} = (0.3,\ 0.2,\ 0.2,\ 0.9,\ 0.3,\ 0.1)$$

The fuzzy clusters F_j of reusable component's that have for 20 components are:

$$F_j = \{C_1, C_2, C_3, C_4, C_5, C_6, C_7, C_8, C_9, C_{10}, C_{11}, C_{12}, C_{13}, C_{14}, C_{15}, C_{16}, C_{17}, C_{18}, C_{19}, C_{20}\}$$

$\mu_{F_1} = (1,\ 0.2,\ 0.1,\ 0.2,\ 0.2,\ 0.1,\ 0.3,\ 0.1,\ 0.1,\ 0.2,\ 0.3,\ 0.2,\ 0.1,\ 0.2,\ 0.1,\ 0.2,\ 0.3,\ 0.4,\ 0.2,\ 0.1)$

$\mu_{F_2} = (0.2,\ 1,\ 0.1,\ 0.1,\ 0.2,\ 1,\ 0.2,\ 0.3,\ 0.2,\ 0.1,\ 0.1,\ 0.2,\ 0.1,\ 0.2,\ 0.1,\ 0.2,\ 0.3,\ 0.4,\ 0.3,\ 0.2)$

$\mu_{F_3} = (0.2,\ 0.1,\ 1,\ 1,\ 0.2,\ 0.3,\ 0.2,\ 0.1,\ 0.2,\ 0.1,\ 0.1,\ 0.2,\ 0.2,\ 0.1,\ 0.2,\ 0.2,\ 0.2,\ 0.1,\ 1,\ 0.2)$

$\mu_{F_4} = (0.2,\ 0.3,\ 0.2,\ 0.1,\ 1,\ 0.2,\ 0.3,\ 0.4,\ 0.1,\ 0.2,\ 1,\ 0.2,\ 0.1,\ 0.3,\ 0.2,\ 0.3,\ 0.2,\ 0.1,\ 0.1,\ 0.2)$

$\mu_{F_5} = (0.1,\ 0.2,\ 0.1,\ 0.4,\ 0.3,\ 0.1,\ 1,\ 0.2,\ 0.2,\ 0.1,\ 0.2,\ 0.1,\ 0.3,\ 0.1,\ 0.2,\ 0.1,\ 1,\ 0.2,\ 0.1,\ 0.1)$

$\mu_{F_6} = (0.2,\ 0.3,\ 0.1,\ 0.2,\ 0.1,\ 0.3,\ 0.1,\ 1,\ 1,\ 1,\ 1,\ 1,\ 0.2,\ 0.1,\ 0.3,\ 0.2,\ 0.3,\ 0.2,\ 0.1,\ 0.2,\ 0.2)$

$\mu_{F_7} = (0.1,\ 0.2,\ 0.2,\ 0.2,\ 0.1,\ 0.3,\ 0.2,\ 0.2,\ 0.1,\ 0.2,\ 0.2,\ 1,\ 0.1,\ 0.2,\ 0.1,\ 0.2,\ 0.1,\ 1,\ 0.1,\ 0.1)$

$\mu_{F_8} = (0.3,\ 0.1,\ 0.2,\ 0.1,\ 0.2,\ 0.1,\ 0.2,\ 0.1,\ 0.2,\ 0.3,\ 0.2,\ 0.2,\ 1,\ 0.1,\ 0.2,\ 0.1,\ 0.2,\ 0.1,\ 0.2,\ 0.1)$

$\mu_{F_9} = (0.1,\ 0.2,\ 0.1,\ 0.3,\ 0.2,\ 0.1,\ 0.2,\ 0.2,\ 0.1,\ 0.2,\ 0.1,\ 0.2,\ 0.1,\ 1,\ 0.3,\ 0.2,\ 1,\ 0.2,\ 0.2,\ 0.1).$

The output class (decision) $D_1 \ldots D_4$ has its membership values as:

$$D = \{C_1, C_2, C_3, C_4, C_5, C_6, C_7, C_8, C_9, C_{10}\}$$

$\mu_{D_1} = (0.0, 0.3, 0.9, 0.6, 0.1, 0.8, 0, 0.3, 0.3, 0.3, 0.1, 0, 0, 0, 0, 0.9, 0.3, 0.9, 0.9, 0.9)$

$\mu_{D_2} = (0.3, 0.9, 0.4, 0.6, 0.2, 0.3, 0.2, 0.9, 0.9, 0.9, 0.2, 0.1, 0.1, 0.2, 0.1, 0.3, 0.9, 0.6, 0.6, 0.4)$

$\mu_{D_3} = (0.2, 0.3, 0.3, 0.3, 0.3, 0.2, 0.9, 0.3, 0.3, 0.3, 0.9, 0.2, 0.4, 0.4, 0.2, 0.2, 0.3, 0.3, 0.3, 0.2)$

$\mu_{D_4} = (0.1, 0.1, 0.1, 0.1, 0.9, 0.1, 0.3, 0.1, 0.1, 0.1, 0.3, 0.9, 0.9, 0.9, 0.9, 0, 0.1, 0, 0, 0)$.

4.8 Conclusions

The integration of CBR and decision theory based on Computational Intelligence techniques has shown its usefulness in the retrieval and selection of reusable software components from a software components repository. Software components are denoted by cases with a set of features, attributes, and relations of a given situation and its associated outcomes. These are taken as inputs to a Decision Support tool that classifies the components as *adaptable to the given situation with membership values for the decisions*. This classification is based on Rough-Fuzzy set theory and the methodology is explained with illustrations. In this novel approach, CBR and DSS (based on Rough-Fuzzy sets) have been applied successfully to the software engineering domain to address the problem of retrieving suitable components for reuse from the case data repository. A software tool called RuFTool is developed as a decision support tool for component retrieval for reuse on Windows platform with MS-ACCESS as the back-end and Visual BASIC as the front-end to this purpose. The use of rough-fuzzy sets increase the likelihood of finding the suitable components for reuse when exact matches are not available or are very few in number.

References

1. Aleti, Aldeida, Buhnova, Barbora, Grunske, Lars, Koziolek, Anne, Meedeniya, Indika: Software architecture optimization methods: a systematic literature review. IEEE Trans. Softw. Eng. **39**(5), 658–683 (2013)
2. Meyer, B.: Reusable Software—The Base Object-Oriented Components Libraries. Hertfordshire: Prentice-Hall, UK, HP27E (1994)
3. Basili, V.R., Rombach, H.D.: Support for comprehensive reuse. Softw Eng. J. **6**(5), 303–316 (1991)
4. Baumer, D., Knoll, R., Linienthal, C., Riehle, D., Zullighoven, H.: Large scale object-oriented software-development in a banking environment: An experience report. In: Cointe P. (eds) ECOOP'96-Object Oriented Programming, 10th European Conference, Linz, Austria, July 1996. Springer-Verlag, New York (1996)
5. Beam, W.R., Palmer, J.D., Sage, A.P.: Systems engineering for software productivity, IEEE Trans. Syst. Man Cybern. SMC-**17**(2), 163–185 (1987)
6. Biggerstaff, T., Perlis, A.: Software Reusability: Vol.1-Concepts and Models, and Vol.2-Applications and Experience. ACM press, New York (1989)

7. Carma, McClure: Adding Reuse to the System Development Process, Englewood Cliffs. Prentice-Hall PTR, NJ (1997)
8. Cox, E.: Fuzzy Systems Handbook (2ed). Academic Press, New York (1999)
9. Damiani, E., Fugini, M.G., Bellettini,C.: A hierarchy-aware approach to faceted classification of object-oriented components, ACM Trans. Softw. Eng. Methodol. **8**(3), 215–262 (1999)
10. Damiani, E., Fugini, M.G.: Fuzzy identification of distributed components, in computational intelligence-theory and applications. In: Reusch, B. (ed) Springer Lecture Notes in Computer Science, vol. 1226, pp. 550–552. Springer, New York, (1997)
11. D'Souza, D., Wills, A.C.: Catalysis: Component and Framework-Based Development, Reading. Addison-Wesley, Massachusetts (1997)
12. Dubois, D., Prade, H.: Rough-fuzzy sets and fuzzy-rough sets. Int. J. Gen Syst. **17**(2–3), 191–209 (1990)
13. Dubois, D., Prade, H.: Putting rough sets and fuzzy sets together. In: Slowinski, R. (ed.) Intelligent Decision Support-Handbook of Applications and Advances of the Rough set theory. Kluwer Academic Publishers, Dordrecht (1992)
14. Frakes, W., Isoda, S.: Success factors of systematic reuse. IEEE Softw. **11**(5), 14–19 (1994)
15. Frakes, W.B., Kang, K.: Software reuse research: status and future. IEEE Trans. Softw. Eng. **31**(7), 529–536 (2005)
16. Gomaa, H.: A reuse oriented approach for structuring and configuring distributing applications. Softw. Eng. J. **8**(2), 61–72 (1993)
17. Jacobson, I., Griss, M., Jonsson, P.: Software Reuse—Architecture, Process and Organization for Business Success. ACM Press, New York (1997)
18. Karlsson, E.A. (ed.): Software Reuse—A Holistic Approach. Wiley, England (1995)
19. Klir, G.J., Folger, T.A.: Fuzzy sets, Uncertainty, and Information. Prentice-Hall Inc., Englewood Cliffs (1988)
20. Kotonya,G., Lock, S, Mariani, J.: Scrapheap software development: lessons learnt from an experiment on opportunistic reuse, IEEE Softw. 68–74 (2011)
21. Koziolek, H.: Performance evaluation of component based systems—a survey. Perform. Eval. **67**(8), 634–658 (2010)
22. Krueger, C.W.: Software reuse. ACM Comput. Surv. **24**(2), 131–183 (1992)
23. Lau, K.-K., Wang, Z.: Software component models. Trans. Softw. Eng. **33**(10), 709–724 (2007)
24. Lillie, C.: Now is the time for a national software repository.In: Proceedings of AIAA on Computing in Aerospace, Baltimore, MD, Oct 1991
25. Meyer, B.: Reusability: the case for object-oriented design. IEEE Softw. **4**(2), 50–63 (1987)
26. Mi, P., Scacchi, W.: A knowledge-based environment for modeling and simulating software engineering processes. IEEE Trans. Knowl. Data Eng. **2**(3), 283–294 (1990)
27. Mili A., Mili, R., Mittermeir R.T.: A survey of software reuse libraries. Ann. Softw. Eng. **5,** (1989)
28. Mili, H., Mili. A., Yacoub, S., Addy, E.: Reuse-based Software Engineering-Techniques, Organisation, and Controls. Wiley, USA (2002)
29. Mili, H., Mili, F., Mili, A.: Reusing software: issues and research directions. IEEE Trans. Softw. Eng. **21**(6), 528–562 (1995)
30. Mili, A., Mili, R., Mittermeir, R.T.: Storing and retrieving software components: a refinement-based system. IEEE Trans. Softw. Eng. **23**(7), 445–460 (1997)
31. Ning J.Q.: A component based software development model. In: Proceedings COMPSAC '96, Seoul, Korea. IEEE Computer Society Press, Los Alamitos, California, pp. 389–394. 1996
32. Pawlak, Z.: Rough sets. Int. J. Comput. Inform. Sci. **11**, 341–356 (1982)
33. Pawlak, Z., Busse, J.G., Slowinsky, R., Ziarko, W.: Rough sets. Commun. ACM **38**(11), 89–95 (1995)
34. Pfleeger, S.L.: Measuring reuse: a cautionary tale. IEEE Softw. **13**(4), 118–125 (1996)
35. Poulin, J.S., Caruso, J.M., Hancock, D.R.: The business case for software reuse. IBM Syst. J. **32**(4), 567–594 (1993)

36. Prieto-Diaz, R.: Status report: software reusability. IEEE Softw. **10**(3), 61–67 (1993)
37. Rada, R., Wang, W., Mili, H., Heger, J., Scherr, W.: Software reuse-from text to hypertext. Softw. Eng. J. **7**(5), 311–321 (1992)
38. Reifer, D.J.: Practical Software Reuse. Wiley, USA (1997)
39. Sametinger, J.: Software engineering with reusable components. Springer, Berlin (1997)
40. Sarkar, M.: Uncertainty-based pattern classification by modular neural networks, Ph.D. thesis, Department of Computer Science and Engineering, Indian Institute of Technology, Madras, India, Oct 1998.
41. Selby, R.W.: Enabling reuse-based software development of large-scale systems. IEEE Trans. Softw. Eng. **31**(6), 418–436 (2005)
42. Sen, A.: The role of opportunism in the software design reuse process. IEEE Trans. Softw. Eng. **23**(7), 418–436 (1997)
43. Standish, T.A.: An essay on software reuse. IEEE Trans. Softw. Eng. **10**(5), 494–497 (1984)
44. Staringer, W.: Constructing applications from reusable components, IEEE Softw. **11**(5), 61–68 (1994)
45. Tirso, J.R., Gregorius, H.: Management of reuse at IBM. IBM Syst. J. **32**(4), 61–68 (1993)
46. Vijay Rao, D.: A unified approach to quantitative lifecycle modeling. Ph.D. Thesis, Department of Computer Science and Automation, Indian Institute of Science, Bangalore, India, 2001
47. Visaggio, G., Process improvement through data reuse, IEEE Software, July 1994, pp. 76-85
48. Wasmund, M.: Implementing critical success factors in software reuse. IBM Syst. J. **32**(4), 595–611 (1993)
49. Weide, B.W., Ogden, W.F., Stuart, H.: Reusable software components, Adv. Comput. **33**, 1–65 (1991)
50. Zadeh, L.A.: Fuzzy sets as a basis for a theory of possibility. Fuzzy Sets Syst. **1**, 3–28 (1978)
51. Zadeh, L.A.: Fuzzy sets. Inf. Control **8**, 338–353 (1965)

Chapter 5
A Soft Computing Approach to Model Human Factors in Air Warfare Simulation System

D. Vijay Rao and Dana Balas-Timar

Abstract With increasing defence budget constraints and environmental safety concerns on employing live exercises for training, there has been a reinforced focus end considerable efforts on designing military training simulators using modelling, simulation, and analysis for operational analyses and training. Air Warfare Simulation System is an agent-oriented virtual warfare simulator that is designed using these concepts for operational analysis and course of action analysis for training. A critical factor that decides the next course of action and hence the results of the simulation is the skill, experience, situation awareness of the pilot in the aircraft cockpit and the pilots' decision making ability in the cockpit. Advances in combat aircraft avionics and onboard automation, information from onboard and ground sensors and satellites poses a threat in terms of information and cognitive overload to the pilot, and triggering conditions that makes decision making a difficult task. Several mathematical models of the pilot behaviour, typically based on control theory, have been proposed in literature. In this work, we describe a novel approach based on soft computing and Computational Intelligence paradigms called ANFIS, a neuro-fuzzy hybridization technique, to model the pilot agent and its behaviour characteristics in the warfare simulator. This emerges as an interesting problem as the decisions made are dynamic and depend upon the actions taken by enemy. We also build a pilots' database that represents the specific cognitive characteristics, skills, training experience, and as factors that affect the pilot's decision making and study its effect on the results obtained from the warfare simulation. We illustrate the methodology with suitable examples and lessons drawn from the virtual air warfare simulator.

D. Vijay Rao (✉)
Institute for Systems Studies and Analyses, Defence Research and Development
Organisation, Metcalfe House, Delhi 110054, India
e-mail: doctor.rao.cs@gmail.com

D. Balas-Timar
Faculty of Educational Sciences, Psychology and Social Sciences,
Aurel Vlaicu University of Arad, Arad, Romania
e-mail: dana@xhouse.ro

V. E. Balas et al. (eds.), *Innovations in Intelligent Machines-5*,
Studies in Computational Intelligence 561, DOI: 10.1007/978-3-662-43370-6_5,
© Springer-Verlag Berlin Heidelberg 2014

5.1 Introduction

With increasing defence budget constraints and environmental safety concerns on employing live exercises for training, there has been a reinforced focus with considerable efforts and investments on designing intelligent military training simulators using modelling, simulation, and analysis for operational analyses and training. Virtual warfare analysis systems constitute an important class of applications that have proved to be an important tool for military system analyses and an inexpensive alternate to live training exercises. Air Warfare Simulation System (AWSS) is a virtual warfare analysis software that has been developed for planning, analysis and evaluating air tasking operations [1, 2]. In the design and development of such applications, modelling the complexity and battle dynamics, assessing, and predicting the outcomes of mission plans quantitatively under various real-world conditions is a very difficult endeavour [3, 4]. However, in most simulator designs, human factors and pilots' decision making plays an important role in determining the outcome of any give situation. These factors are considered and hence the effects of dynamic decisions based on enemy actions are not incorporated. This forms the necessary basis for building a persona of pilot agents whose behaviour and plausible decisions are identified and simulated in the air warfare simulator. An agent-based approach to designing the air warfare simulator is discussed in Sect. 5.2, mathematical models for describing pilot behaviour is described in Sect. 5.3, the soft computing modelling approach to model the pilot behaviour proposed in this chapter using adaptive neuro-fuzzy inference system (ANFIS) is described in Sect. 5.4. This is followed by two conflict situations as a case study of human factors in decision making and concluding with discussion of the results and future work. This example case study is based on pilot's skill data and decisions taken in various conflict situations and the data collected in simulated environments is used as a training data set for the pilot agent in the air warfare simulation system.

5.2 An Agent-Based Architecture to Design Military Training Operations

Agent-oriented systems' development aim to simplify the construction of complex systems by introducing a natural abstraction layer on top of the object-oriented paradigm composed of autonomous interacting actors [37–39]. It has emerged as a powerful modeling technique that is more realistic for today's dynamic warfare scenarios than the traditional models which were deterministic, stochastic or based on differential equations [24–29, 44–51]. These approaches provide a very simple and intuitive framework for modeling warfare and are very limited when it comes to representing the complex interactions of real-world combat because of their high degree of aggregation, multi-resolution modeling and varying attrition rate

Fig. 5.1 Agent architecture for AWSS

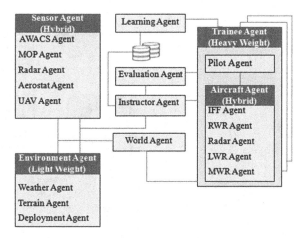

factors. The effects of random individual agent behavior and of the resulting interactions of agents are phenomenon that traditional equation-based models simply cannot capture. The traditional approaches of computers to warfare simulation used algorithms that aggregated the forces on each side, such as differential equations or game theory, effectively modeling the entire battle space with a single process [23, 34]. These mathematical theories treat the opposing sides as aggregates, and do not consider the detailed interactions of individual entities.

Agent-based models [38, 39] give each entity its own thread of execution, mimicking the real-world entities that affect military operations [3, 42, 43, 52–54]. The pilot agent is a heavy-weight agent that determines the current environmental conditions over the area of operation selected for the mission. We define a heavy-weight agent as one that has its own decision-making ability and a built-in capability to select the next course of action, based on the inference from a given *situation* [40, 41]. The *situation* is generated from information generated by other agents and the agent takes a reasoned decision and action (Figs. 5.1, 5.2).

Conventionally, the pilot aspects have either been ignored or implemented as a set of broad rules that do not have a bearing on the results of the war simulation. This has led to a large number of missions planned by the trainees that are statistically correct but not accepted realistically as each trainee had a different story to add to the mission he flew. The newer form of sensors and their integration in a networked environment changes the entire warfare scenario to the extent that the results obtained by the training simulator were not realistic. These inputs provided us an impetus to consider the human factors in decision making and explored the various inputs obtained from the trainee pilots [31–33].

The *pilot agent* receives information from other agents such as weather, terrain and deployment agent and provides an information service to the world agent after its own process of reasoning and actions. This information is then used by other agents such as Manual Observation Post (MOP), Unmanned Air Vehicle (UAV), Identification Friend/Foe (IFF), Radar Warning Receiver (RWR), Missile Warning Receiver (MWR), Laser Warning Receiver (LWR), *Mission Planning, Sensor*

Fig. 5.2 Design of the pilot agent in AWSS

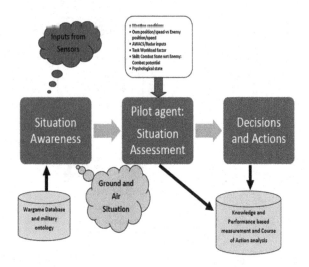

Performance, Target Acquisition and *Damage Assessment and Computation.* All these agents add to the information update in simulated real time and generate the situation awareness for the pilot agent. A number of approaches have been used to model the pilot depending upon the goal of study. In this chapter, we designed the pilot agent using ANFIS, a neuro-fuzzy hybridization technique that takes all these factors as fuzzy inputs and generates the combat potential utility [19] for a given situation. This is used to generate the dynamic decisions and the courses of actions for a given situation of conflict as opposed to the statistical approach to generating the rules for decision making. The results obtained from this approach to model the pilot agent has proved to be better at generating realistic effects of pilot decisions and human factors involved in decision making [20–22]. The analysis has brought out the nuances of decisions at every stage of the wargame and also the effects of various factors are analysed for identifying the causes of the predicted results [1].

5.3 Mathematical Modelling of Pilot Behavior

Several models to represent the pilot's cognitive behavior have been proposed in literature. When modeled from a control systems perspective, the pilot can be represented by a number of complex block diagrams which describe all the possible factors affecting human behavior [7, 55–60]. It is not possible, however, to create one universal cognitive model fully describing the dynamic human behavior in all flying/combat situations.

One possible model of human behavior dynamics is shown in the block diagram in Fig. 5.3. It represents a simplified and concise model of the pilot's behavior. There are three mutually interacting sub-systems. The input-sensors are the pilot's sensory organs, from where the detected information goes into the central nervous

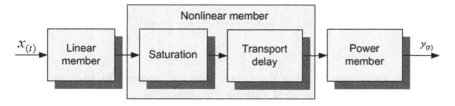

Fig. 5.3 A simplified human behavior model

system. The average speed of emotion transmission is in the range of 5–125 m/s. In an automated control system this transmission feature can be represented by a transport delay. The response times mainly depends on the level of the pilot's internal stress, the pilot's actual conditions and perhaps also on some other external factors such as information overload conditions, attention and decision making. Sensory organ features are in real life represented by a sensitivity level, adaptation ability and the ability to mutually cooperate. After processing the received signal a command to hand or leg muscles is sent to adjust the elevator, aileron and rudder deflections. For maintaining the requested flight parameters the pilot uses three different types of regulators [55, 58]:

- Predictive regulator: Keeps the required flight mode based on the pilot's visual and sensory perception of the flight.
- Feedback regulator: Created based on the visual and sensory perception of the required flight mode.
- Precognitive regulator: Recall the learnt maneuver from memory, i.e. a clear sequence of elevator, aileron and rudder deflections making the required aircraft movement.

It is a difficult endeavor to describe human/pilot behavior mathematically. It is difficult to completely list all the physiological and cognitive processes of the human brain and therefore it is not possible to create a comprehensive list describing the human thinking processes upon which pilot behavior is based. A human—as a pilot—is able to adapt and fly various types of aircraft after a certain amount of training. A human can also manage complex situations by adapting his behavior based on current conditions, and is capable of changing his strategy and tactics based on visual input information. The decision making process and choice of future action is, more or less, individual, especially in emergency situations.

When analyzing any aircraft control with human behavior, it is essential to take into account that all the human features are time variables and dependent on the actual pilot condition, psychological state, tiredness and ability to adapt to a new situation. This is all affected by long-term habits, education, training. To create a mathematical model of a human in such a moment is not easy. For modeling human behavior a linear model is often being used (which is not quite correct for example regarding output value limitations) with a transport delay defined by a transfer function as follows [55, 56, 59, 60]:

$$F_{(s)} = \frac{Y_{(s)}}{X_{(s)}} = K \frac{(T_3 s + 1)}{(T_1 s + 1)(T_2 s + 1)} e^{-\tau s}, \qquad (5.1)$$

where:

K *Pilot Gain*—representing the pilot's ability to respond to an error in the magnitude of a controlled variable. Increasing of force on the steers in relation to their deflection (from 0.1 to 100)

T_1 *Lag Time Constant*—describes the ease with which the pilot generates the required input i.e. reaction ability to rate of change of input signal (0.1–0.4 s)

T_2 *Neuromuscular Lag Time Constant*—represents the time constant associated with contraction of the muscles through which the control input is applied by the pilot. The dynamics properties of the pilot power member's components (0.05–0.2 s)

T_3 *Lead Time Constant*—reflecting the pilot's ability to predict a control input (0–2 s)

τ Represents a pure time delay describing the period between the decision to change a control input and the change starting to occur (0.1 to 0.3 s)

This shape of the transfer function is based on the assumptions and can be applied in cases where the pilot behaves as a linear member. In the real control loop non-linear elements are modeled as in the case of pilot—aircraft system. In the literature [7], for cases where the nonlinearity of actuator is also modeled, extended shape of above mentioned transfer function in the shape:

$$F_{(s)} = \frac{Y_{(s)}}{X_{(s)}} = K \frac{(T_3 s + 1)}{(T_1 s + 1)(T_2 s + 1)} e^{-\tau s} + remnant\ function \qquad (5.2)$$

The design of the remnant function is a tedious procedure as it attempts to represent the on-linear component of pilot behavior. This is the primary contribution where it denotes the pilot's ability to learn and adapt to new situations resulting in non-linear and non-steady behavior. The secondary contribution comes from other factors depending on the experimental setup and experimentally injected noise that affect pilot's response caused to other inputs. However, a careful selection of the pilot model and tasks can help minimize remnant effect [2]. In fact, the human operator does not perform controlling activities according to a linear model, but his control efforts are always loaded by negative effects of nonlinear elements such a hysteresis, dead zone, saturation or nonlinear variable gain. It is difficult to identify not only those elements but also include or placed elements into the regulation circuit which has multiple feedback. Many researchers use a transfer function for pilot compensation response as shown in Eq. (5.1). This equation was first published by the English scientist Arnold Tustnin who was studying the characteristics of a human regulator with manual feedback control. Similar physiological analysis of time constants in the aforementioned human regulator transfer function was done in the 1960s by the American scientist

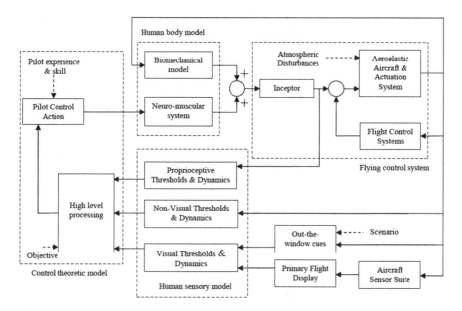

Fig. 5.4 Block diagram representing the pilot-vehicle system under manual control [7]

McRuer for autopilot models. There are several studies done in describing scientists assigning individual time constants to physiological processes. However, there are many opponents stating that this approach is not correct as neuro-motive functions and central nervous system functions are mixed together.

Figure 5.4 shows the key control system blocks along with the pilot and autopilot sharing control of the aircraft. The pilot is the main controlling element who knows the flight task. All the automated functions are adjusted to this condition, including the pilot control options and Flight Control Systems. The pilot's response is, based on the pilot's flight experience and skills, processed at the subconscious level using higher brain functions. This information is then transferred via the neuromuscular system creating a physical response. The pilot then maneuvers the aircraft using appropriate aircraft elevator deflection. The pilot senses the resultant aircraft movement through many different senses observed as feedback to the pilot. This feedback is then processed again by the neuromuscular system and the aircraft elevator deflections are adjusted accordingly. The Flight Control System significantly simplifies the flying process for the pilot by checking and adjusting a whole range of flight parameters from damping fast oscillations to flatter damping.

One of the possible methods of calculating parameters of pilot behavior model transfer function is using MATLAB using System Identification Toolbox [18]. The System Identification Toolbox helps in designing mathematical models of dynamic systems from measured input and output data. The toolbox performs grey-box

Fig. 5.5 MATLAB System
identification toolbox and
process functions

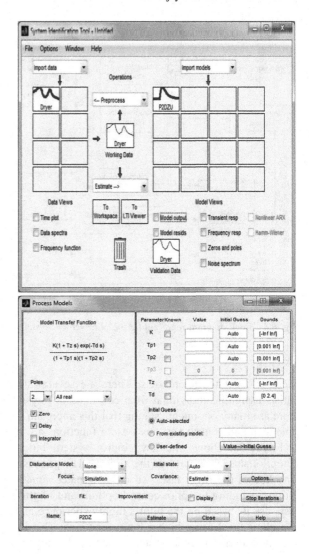

system identification for estimating parameters of a user-defined model. The input
data consisted of the measured changes of altitude depending on control yoke
deviation. The output data consisted of measured control yoke deviation in the
longitudinal direction. The toolbox uses a graphical user interface (GUI) Fig. 5.5,
which facilitates work with organization of data and models. It is possible to use
time-domain and frequency-domain input-output data to identify continuous-time
and discrete-time transfer functions, process models, and state-space models. For
model views the transient response, frequency response, zeros and poles and noise
spectrum is used.

5.4 A Neuro-Fuzzy Hybridization Approach to Model the Pilot Agent in AWSS

Pilot models have been implemented using several approaches depending upon the purpose for which it has been modeled. One of the earliest approaches was to implement using transfer functions in control theory with feedback to model the various operations of the pilot [7]. Recent works using this approach used the Cessna aircraft to capture the data of trainees [55, 57, 58]. As part of our research work, we implemented the pilot agent as a set of decision tree rules that were statistically obtained from a large set of simulated data. These rules were found to be statistically valid for a large scale warfare simulation that was typically performed as part of the strategic warfare analysis. However, this approach did not divulge the various nuances of reasoning, inference and decision making which has a profound bearing on the outcomes of the warfare. In this chapter, we discuss the pilot agent as modeled by the ANFIS and taking into various cognitive factors of the pilot's decision making processes and situation assessment.

5.4.1 ANFIS

Both neural networks and fuzzy systems are dynamic, parallel processing systems that estimate input–output functions [9, 12–14, 16]. They estimate a function without any mathematical model and learn from experience with sample data. It has also been proven that (1) any rule-based fuzzy system may be approximated by a neural net and (2) any neural net (feed-forward, multi-layered) may be approximated by a rule-based fuzzy system. Fuzzy systems can be broadly categorized into two families. The first includes linguistic models based on collections of IF—THEN rules, whose antecedents and consequents utilize fuzzy values. The Mamdani model falls in this group where the knowledge is represented as it is shown in the following expression [5, 30].

$$R^i : If\ X_1\ is\ A_1^i\ and\ X_2\ is\ A_2^i. \ldots \ldots\ and\ X_n\ is\ A_m^i, then\ y^i\ is\ B^i$$

The second category, which is used to model the Weather prediction problem, is the Sugeno-type and it uses a rule structure that has fuzzy antecedent and functional consequent parts. This can be viewed as the expansion of piece-wise linear partition represented as shown in the rule below.

$$R^i : If\ X_1\ is\ A_1^i\ and\ X_2\ is\ A_2^i. \ldots \ldots\ and\ X_n\ is\ A_m^i, then$$
$$y^i = a_0^1 + a_1^i X_1 + \ldots .. + a_n^i X_n$$

$$\tilde{A} \cap \tilde{B} = \left\{ \left(x, \mu_{\tilde{A} \cap \tilde{B}}(x) \right) | \mu_{\tilde{A} \cap \tilde{B}}(x) = \mu_{\tilde{A}}(x)^\wedge \mu_{\tilde{B}}(x) = \min \left(\mu_{\tilde{A}}(x), \mu_{\tilde{B}}(x) \right) \right\} \quad (5.3)$$

The conjunction "and" Operation between fuzzy sets known as *Linguistics*, for the implementation of the Mamdani rules is done by employing special Fuzzy Operators called T-Norms [6, 10–13]. The ANFIS uses by default the Minimum T-Norm which is the case here and it can be seen in the Eq. 5.1. The approach approximates a nonlinear system with a combination of several linear systems, by decomposing the whole input space into several partial fuzzy spaces and representing each output space with a linear equation. Such models are capable of representing both qualitative and quantitative information and allow relatively easier application of powerful learning techniques for their identification from data. They are capable of approximating any continuous real-valued function on a compact set to any degree of accuracy [9, 10]. This type of knowledge representation does not allow the output variables to be described in linguistic terms and the parameter optimization is carried out iteratively using a nonlinear optimization method.

Fuzzy systems exhibit both symbolic and numeric features. Neuro-fuzzy computing [8, 11, 12] is a judicious integration of the merits of neural and fuzzy approaches, enables one to build more intelligent decision-making systems. Neuro-fuzzy hybridization is done broadly in two ways: a neural network equipped with the capability of handling fuzzy information (termed fuzzy-neural network) and a fuzzy system augmented by neural networks to enhance some of its characteristics like flexibility, speed, and adaptability (termed neural-fuzzy system). ANFIS is an adaptive network that is functionally equivalent to a fuzzy inference system and referred to in literature as "adaptive network based fuzzy inference system" or "adaptive neuro-fuzzy inference system" (Fig. 5.6) [9, 10, 13, 14]. In the ANFIS model, crisp input series are converted to fuzzy inputs by developing triangular, trapezoidal and sigmoid membership functions for each input series. These fuzzy inputs are processed through a network of transfer functions at the nodes of different layers of the network to obtain fuzzy outputs with linear membership functions that are combined to obtain a single crisp output the course of action suggested by the pilot agent, as the ANFIS method permits only one output in the model. The following Eqs. (5.4) (5.5) (5.6) correspond to triangular, trapezoidal and sigmoid membership functions (Figs. 5.7, 5.8a, b).

$$
\mu_s(\mathbf{X}) = \begin{cases} 0 & \text{if } X < a \\ (X - a)/(c - a) & \text{if } X \in [a, c) \\ (b - X)/(b - c) & \text{if } X \in [c, b] \\ 0 & \text{if } X > b \end{cases} \tag{5.4}
$$

$$
\mu_s(\mathbf{X}) = \begin{cases} 0, & \text{if } X \le a \\ (X - a)/(m - a) & \text{if } X \in [a, m) \\ 1, & \text{if } X \in [m, n) \\ (b - X)/(b - n) & \text{if } X \in (n, b) \\ 0, & \text{if } X \ge b \end{cases} \tag{5.5}
$$

Fig. 5.6 Pilot agent's
architecture and behaviors
(decisions) modeled in
AWSS

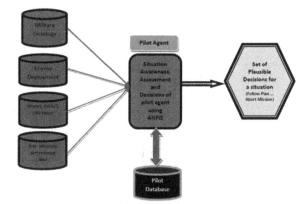

Fig. 5.7 ANFIS architecture
to design the Pilot agent

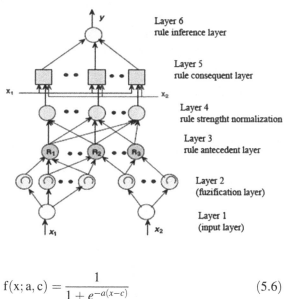

$$f(x; a, c) = \frac{1}{1 + e^{-a(x-c)}} \qquad (5.6)$$

5.4.2 Modeling the Human Factors and Situation Awareness of Pilot Agent

Situation Awareness is the perception of the elements in the environment within a
volume of time and space, the comprehension of their meaning, and the projection
of their status in the near future [15, 35, 36]. Situation awareness (SA) and human
factors modeling is a predominant concern in system operation, and dynamic

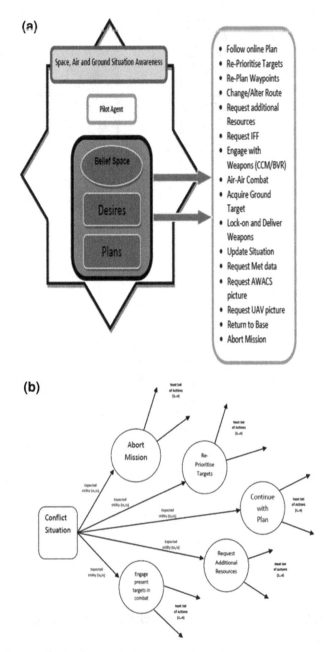

Fig. 5.8 **a**, **b** The role of pilot agent in decision making and course of action analysis in AWSS

decision making in an air warfare scenario. Maintaining a high level of SA is one of the most critical and challenging jobs of the combat pilot in these critical situations. Problems with SA were found to be the leading causal factors in a review of military aviation mishaps [15]. In a study of accidents among major airlines, 88 % of those involving human error could be attributed to problems of SA as opposed to problems with decision making or flying skills [15]. Due to the important role that SA plays in the combat pilot's decision making processes, we designed a pilot's database that captures the attributes representing the flying skills of pilots (from clinical/experiment tests) and these form an input to the pilots' decision making during the air warfare simulation. A study of the errors in SA are summarized as follows: Workload/distraction (86 %), Communications/coordination (74 %), improper procedures (54 %), time pressures (45 %), equipment problems (43 %), weather (32 %), unfamiliarity (31 %), fatigue (18 %), night conditions (12 %), emotion (7 %) and other factors (37 %).

Loss of Level 1 SA: Failure to correctly perceive the situation: 76 %
Loss of Level 2 SA: Failure to correctly comprehend the situation: 20 %
Loss of Level 3 SA: Failure to correctly project situation: 4 %

Several methods of testing situation awareness have been documented [36], notably the knowledge-based assessment and performance-based assessments. Several complex techniques exist which attempt to determine or model the pilot's knowledge of the situation at different times throughout the simulation runs. For example, the Situation Awareness Global Assessment Technique (SAGAT) freezes the simulator screens at random times during the runs, and queries the subjects about their knowledge of the environment. Such an approach has been designed to test the knowledge based measurement using the military ontology. This has been discussed in [7]. This knowledge can be at several levels of cognition, from the most basic of facts to complicated predictions of future states. Several causal factors affect the actions of the subject, as shown in Fig. 5.1. Comparing knowledge-based and performance-based techniques of evaluating situation awareness, we find they take measurements at different points in the process of user cognition.

The scenario in AWSS considered these situations and used the pilot's database depicting the cognitive/behavioral attributes using ANFIS in order to generate the appropriate decisions/actions of the pilots. Each decision of the pilot during the conflict resolution phase of the simulator design asks the pilot for an action, and this leads to the next course of action. The entire lesson plan thus is represented as a network of decision nodes and actions. At the end of the training lesson, analysis of the pilot's decisions and their actions are done by knowledge based measurement and performance based measurements. In the following section, we consider two conflict situations that demand an explicit action by the pilot under training.

5.5 Design of Conflict Situations and Discussion of Results

In the design of training simulators, lessons plans are prepared with an inherent conflict situation that needs to be resolved and arriving at a decision and taking the appropriate action. In this process measurements are taken to evaluate the quality of decisions and lessons learnt. Two measurement based approaches based on the knowledge base and cognitive skills and performance-based measurement of assessing situation awareness were considered to study the effects of pilot decision making process and the end results of the mission effectiveness. Two conflict situations are highlighted in the following scenarios and the decisions/actions of the pilot are recorded.

Situation 1: In an air combat simulation where two opposing combat aircraft are engaging to kill, different pilots take different actions that are dependent upon the situation assessment and also on the actions performed by the opposing force. We simulate these situations with the characteristic profiles (persona) of two different pilots P1 and P2 that are stored in the pilot database.

Situation 2: When the combat pilot of a multi-role aircraft encounters bad weather and is unable to locate the target, various decisions can be taken based upon the pilots experience and ground picture that he is familiar with. Situation awareness and assessment of the pilot and makes different pilots to take different decisions that lead to different mission effectiveness (Table 5.1).

The results of both these situation analysis, decisions taken by the two pilots, and the mission effectiveness are summarized in Tables 5.2 and 5.3. A summary of the various factors that are considered to evaluate a combat effectiveness of a pilot are shown in Table 5.1. Several clinical psychometric tests used to measure all the intangible human factors are stored in the pilot's database. However, for this study, we considered the factors shown in Table 5.2.

The factors identified in Table 5.2 are representative of the two pilots P1 and P2, who differ mainly in *Information Processing* and decision making, *Risk taking* and *Reaction to stress* which are typically identified personality traits. Data collected using clinical and psychometric tests for all the pilots are stored in the Pilot's database. These attribute values from the pilot's database (Fig. 5.9) are fuzzified and used to determine the pilot's personality as one of the inputs to the ANFIS tool. The other inputs that are used for computing the Combat Utility factor are: Type of combat aircraft, Mission commander (pilot and skills as in Table 5.2), number of sorties flown on the day of mission, expected enemy air/ground defence threat, situation awareness of the pilot, entropy or uncertainty of the sensor fused information to the pilot, combat potential ratio (of own and enemy resources), pilot's subjective mental workload measured using the NASA-TLX measurement scale, Combat Utility factor that measures the combat utility of the pilot in taking a decision and following a course of action as depicted in Fig. 5.8a, b. These input factors of pilots play an important role in decision making which in turn affects the overall mission success. The output obtained from the ANFIS system is a measure called Combat Utility factor that depicts the utility of the pilot in achieving his

Table 5.1 Combat effectiveness factors for Pilots

---

1. Biographical data
   a. Life inventory
   b. Academic history
   c. Military history
   d. Military rank
2. Risk taking
   a. Willingness to take calculated risks
3. Reactions to stress
   a. Performance under stress
   b. Emotional control
   c. Ability to withstand psychological stress
   d. Anxiety
4. Sensory-motor abilities
   a. Visual perception
   b. Motor coordination
   c. Spatial coordination
   d. Spatial-perceptual ability
   e. Perceptual speed
5. Aptitude
   a. Pilot composites
   b. Non-pilot composites
   c. General aptitude (Intelligence)
   d. Numerical skills
   e. Verbal skills
   f. Mechanical skills
   g. Flight aptitude
6. Personality
   a. Aggressiveness
   b. Self confidence
   c. Mental health
   d. Consideration for others
   e. Personality style
   f. Courage
7. Personality-leadership
   a. Responsibility for men in combat
   b. Physical and combat leadership
   c. Adinistrative skills
   d. Mlitary bearing
8. Social factors
   a. Teamwork
   b. Sociability
   c. Group loyalty
   d. Interpersonal rating

(continued)

Table 5.1 (continued)

9. Motivation
  a. Determination/desire
  b. Self discipline
  c. Satisfaction
10. Medical/physiological
  a. Good physical health
  b. Endurance
  c. Physical aptitude
  d. Ability to withstand psychological stress
11. Decision making/information processing
  a. Selective attention
  b. Decision time
  c. Quality of combat decisions
  d. Alertness
  e. Integrative decisions
12. Aviator skills, knowledge and tasks
  a. Equipment knowledge
  b. Flight skills
  c. Instrument reading
  d. Aerial gunnery
  e. BVR/CCM launch
  f. Stealth
  g. Air combat

objective. In a similar vein, we also compute the Combat Utility factor for the enemy.

Based on the values obtained from the ANFIS tool, the pilot agent takes a rational decision that is also depicted in Table 5.3. Decisions that are obtained from the ANFIS are used as the pilot's decision and the mission effectiveness is computed in AWSS. The ANFIS tool is developed in Matlab [18] and has an interface with a Visual Basic application of the AWSS. Some typical results that are obtained using the ANFIS tool and AWSS that consider the human factors, situational awareness and assessment, and pilot skills are summarized in Table 5.3.

Some typical results that are obtained using the ANFIS tool and AWSS that consider the human factors, situational awareness and assessment, and pilot skills are summarized in Table 5.3. The situation 1 depicted as an air to air combat, commanded by pilot P1 with Very High Situation awareness and High uncertainty in this overloaded information has Low mental workload (NASA-TLX) and decides to "Re-Prioritize targets" thereby choosing the optimal combat utility factor to obtain a Mission success of 8.7; whereas another pilot P2 with High mental workload decides to "Request Additional Resources" as this option seemed to yield a high combat utility for him and achieving a Mission success of 4.3.

Table 5.2 Pilot's Attributes considered in the ANFIS

Pilot Id	Personality type	Risk taking	Info processing and risk taking	Aviator skills/ experience	Firing skills/ experience	Sensor–motor abilities	Personality– leadership	Motivation	Reactions to stress	Physiological/ medical health
P1	A	High	High	Very high	Very high	High	Excellent	High	Composed	Cat 1
P2	B	Low	Low	High	High	High	Excellent	High	Stressed	Cat 1

Table 5.3 Pilot attributes and decisions to determine the mission success factor in conflict situations

MissionID	Combat aircraft	Mission comdr	Sorties flown/day	Enemy air/ground defence threat	Situation awareness	Entropy	Information overload	Combat potential ratio (MAUT) (Own:En)	Combat utility ratio (ANFIS) (Own:En)	Subjective mental workload NASA–TLX	Situational Decision of pilot (AWSS)	Mission success factor
#001 Sit 1	Multi-role	P1	3	High	Very high	High	High	Static: 1:0.78 Dyn: 1:1.3	1: 1.4	Low	Re-prioritize targets	8.7
#001* Sit 1	Multi-role	P2	3	High	Very high	High	High	Static: 1:0.78 Dyn: 1:1.3	1:1.4	High	Request additional resources	4.3
#002 Sit 2	Multi-role	P1	2	Very low	Low	Low	Low	Static: 1:0.44 Dyn: 1:0.65	1:-0.3	Low	Lock-on and deliver weapons	7.6
#002* Sit 2	Multi-role	P2	2	Very low	Low	Low	Low	Static: 1:0.44 Dyn: 1:0.65	1:-0.3	High	Look for secondary targets	3.2

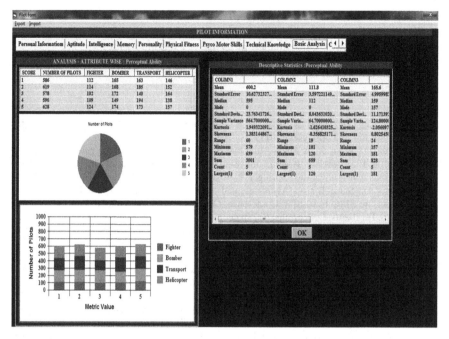

Fig. 5.9 Pilot database used for decision making

Similarly, in Situation 2, which depicts an air-to-ground targeting scenario with bad weather, pilot P1 decides to "Lock-on and Deliver Weapons" taking the appropriate risk and obtaining a mission success of 7.6, whereas pilot P2 decides to "Look for alternate targets" thereby achieving a mission success of 3.2.

5.6 Conclusions and Future Work

A novel approach to design the pilot agent using ANFIS is presented in this research chapter. More specifically, a neuro-fuzzy hybridization technique is employed to model the pilot skills factors and the operations of the pilot agent in a virtual warfare analysis system called AWSS that is designed using an agent-based architecture. The system is applied to compare the results obtained by considering the pilot decision factors to obtain a useful measure called Combat Utility in combat simulation exercises to take the next course of actions. The results that are predicted by the pilot agent after training exercises and the rules that are generated to predict the Mission_Success_Factor are found to be very satisfactory in predicting the mission's performance in the presence of different situations and pilots' persona with different skill attributes. This concept introduces a new approach to introduce human factors and their effects in the AWSS by using ANFIS as the reasoning and inference system. Future research efforts include working on ways for optimizing the rules for specific combat scenario and also on the improvement of the overall system's performance in terms of execution time.

References

1. Vijay Rao, D.: The Design of Air Warfare Simulation System. Technical report, Institute for Systems Studies and Analyses (2011)
2. Vijay Rao, D., Kaur, J.: A Fuzzy Rule-based approach to design game rules in a mission planning and evaluation system, In: 6th IFIP Conference on Artificial Intelligence Applications and Innovations, Springer, New York (2010)
3. Vijay Rao, D., Saha, B.: An Agent oriented Approach to Developing Intelligent Training Simulators. In: SISO Euro-SIW Conference, Ontario, Canada (June 2010)
4. Vijay Rao, D., Iliadis, L. Spartalis, S. Papaleonidas, A.: Modelling Environmental factors and effects in virtual warfare simulators by using a multi agent approach. Int. J. Artif. Intell. 9(A12), pp. 172–185 (2012)
5. Vijay Rao, D. Iliadis, L. Spartalis, S.: A Neuro-Fuzzy Hybridization Approach to Model Weather Operations in a Virtual Warfare Analysis System". In: Proceedings of the 12th EANN (Engineering Applications of Neural Networks). LNCS AICT, vol. 363(1), pp. 111–121. Springer (2011)
6. Vijay Rao, D. Ravi, S. Iliadis, L. Sarma, V.V.S.: "An Ontology based approach to designing adaptive lesson plans in military training simulators". In: Proceedings of the 13th EANN (Engineering Applications of Neural Networks), LNCS AICT, vol. 363(1), pp. 81–93. Springer (2012)
7. McRuer, D.T. Krendel, E.S.: Mathematical models of human pilot behavior, AGARD monograph No. 188, NATO advisory group on Aerospace research and development. (AGARD), London (1974)
8. Banks, J.: Handbook of Simulation: Principles, Methodology, Advances, Applications, and Practice. Wiley, New York (1998)
9. Cox, E.: The Fuzzy Systems Handbook, 2nd edn. Academic Press, New York (1999)
10. Taher, J. Zomaya, A.Y.: In: Zomaya, A.Y. (Ed.)Artificial Neural Networks in Handbook of Nature-Inspired and Innovative Computing, Integrating Classical Models with Emerging Technologies, pp. 147–186, Springer USA (2006)
11. Jang, J.S.R.: ANFIS: Adaptive network-based fuzzy inference systems. IEEE Trans. Syst. Man Cybern. 23(3), 665–685 (May/June 1993)
12. Jang, J.S.R., Sun, C.T., Mizutani, E.: Neuro-Fuzzy and Soft Computing: A Computational approach to Learning and Machine Intelligence. Prentice Hall, Mahwah (1997)
13. Mitra, S., Yoichi, Hayashi: Neuro-fuzzy rule generation: survey in soft computing framework. IEEE Trans. Neural Netw. 11(3), 748–768 (2000)
14. Mendel, J.M.: Uncertain Rule-Based Fuzzy Logic Systems-Introduction and New Directions. Prentice Hall PTR, Upper Saddle River, USA (2001)
15. Endsley, M.R., Garland, D.J. (eds.): Situation Awareness Analysis and Measurement. Lawrence Erlbaum Associates Publishers, Mahwah (2000)
16. Karray, F.O., De Silva, C.: Soft Computing and Intelligent Systems-Theory. Tools and Applications, Pearson-Addison Wesley, England (2004)
17. MATLAB Fuzzy Logic Toolbox Mathworks http://www.mathworks.com/help/fuzzy/
18. Nunn,W.R. Oberle, R.A.: Evaluating Air Combat Maneuvering Engagements, Methodology, Center for Naval Analyses, 1402 vol. I Wilson Boulevard, Arlington, Virginia 22209, USA (1976)
19. Parrott, E.: Combat Performance Advantage: Method of Evaluating Air Combat Performance Effectiveness, Aerodynamics and Performance Branch, Technical report ASD-TR-78-

364,Aeronautical Systems Division, Air Force Systems Command, Wright-Patterson AFB, Ohio 45433, USA (1978)

20. Triantaphyllou, E. Mann, S.H.: An Examination of the Effectiveness of Multi-Dimensional Decision-Making Methods: A Decision-Making Paradox, Decision Support Systems vol. 5, 303–312, North Holland (1989)

21. Jaiswal, N.K.: Military Operations Research: Quantitative Decision Making. Kluwer Academic Publishers, Boston (1997)

22. Markushostmanna, Bernauer, Hans-Joachimmosler, T.: Peterreichert, Bernhardtruffer, multi-attribute value theory as a framework for conflict resolution in river Rehabilitation. J. Multi-Crit Decis. Anal. **13**, 91–102 (2005)

23. Tolk, A.: Engineering Principles of Combat Modeling and Distributed Simulation. Wiley, USA (2012)

24. Rodin, E.Y., Amin, S.M.: Maneuver prediction in air combat via artificial neural networks. Comput. Math. Appl. **24**, 95–112 (1992)

25. Gorman, P.R., Sejnowski, T.J.: Analysis of hidden units in a layered network trained to classify sonar targets. Neural Netw. **1**, 75–89 (1988)

26. N.H. Farhat, S. Miyahara, K.S. Lee, Optical Analog of Two-Dimensional Neural Networks and their Applications in Recognition of Radar Targets, pp. 146–152. American Institute of Physics, New York (1986)

27. Meier, C.D. Stenerson, R.O.: Recognition networks for tactical air combat maneuvers, In: Proceedings of the 4th Annual AAAI Conference, Dayton, OH (Oct. 1988)

28. Mitchell, R.R.: Expert systems and air-combat simulation. AI Expert **4**(9), pp. 38–43 (Sept. 1989)

29. Morgan, A.J.: Predicting the behaviour of dynamic systems with qualitative vectors. In: Hallam, J. Mellish C. (eds.) Advances in Artificial Intelligence, pp. 81–95. Wiley, New York (1988)

30. Tran, C. Abraham, A. Jain, L.: Adaptation of a Mamdani Fuzzy Inference System Using Neuro-genetic Approach for Tactical Air Combat Decision Support System, AI 2002: Advances in Artificial Intelligence (2557). In: Proceedings 15th Australian Joint Conference on Artificial Intelligence Canberra, pp. 672–680 Springer, Australia 2–6 Dec 2002

31. Heinze, C.: Modelling Intention Recognition for Intelligent Agent Systems, DSTO-RR-0286, DSTO Systems Sciences Laboratory, Edinburgh, South Australia, Australia 5111, (ADA430005), (2004)

32. Wooldridge, M., Jennings, N.R.: Intelligent agents:theory and practice. Knowl. Eng. Rev. **10**(12), 115–152 (1995)

33. Bratman, Michael E.: Intention, Plans, and Practical Reasoning. Harvard University Press, Cambridge (1987)

34. Rao A.S. Georgeff, M.P. :An Abstract Architecture for Rational Agents. In: Rich, C. Swartout, W. Nebel, B. (eds.) Proceedings of the Third International Conference on Principles of Knowledge Representation and Reasoning (KR'92), Morgan Kaufmann Publishers, San Francisco, CA, USA (1992)

35. Endsley, M.: Toward a theory of situation awareness in dynamic systems. Hum. Factors **31**(l), 32–64 (1995)

36. Endsley, M.: The role of situation awareness in naturalistic decision making. In: Zsambok C.E. Klein, G. (eds.) Naturalistic Decision Making. Lawrence Erlbaum Associates Publishers, USA (1997)

37. McManus, J.W. Goodrich, K.H. Artificial Intelligence Based Tactical Guidance For Fighter Aircraft. In: AIAA Guidance, Navigation, and Control Conference 20–22 August 1990, Portland, Oregon, USA (1990)

38. Heinze, C. Smith, B. Cross M.: Thinking quickly: agents for modeling air warfare. In Proceedings of the 9th Australian Joint Conference on Artificial Intelligence (AP98), Brisbane, Australia (1998)

39. Rao, A.S. Murray, G.: Multi-agent mental-state recognition and its application to air-combat modeling. In: Proceedings of the 13th International Workshopon Distributed Artificial Intelligence (DAI-94), pp. 283–304. Seattle, WA, USA (1994)

40. Rao, A.: A unified view of plans as recipes. Technical Report 77, Australian Artificial Intelligence Institute, Melbourne, Australia, August (1997)
41. Rao, A.S.: Means-end plan recognition—towards a theory of reactive recognition. In Doyle, J. Sandewall, E. Torasso, P. (eds.) Proceedings of 4th International Conference on Principles of Knowledge Representation and Reasoning, pp. 497–508, Bonn, FRG, May Morgan Kaufmann Publishers; San Francisco, CA, USA (1994)
42. Rao, A.S. Georgeff, M.P.: Modeling rational agents within a bdi-architecture. In: Second International Conference on Principles of Knowledge Representation and Reasoning, San Mateo, CA, (1991)
43. Rao, A.S. Georgeff, M.P.: BDI-agents: from theory to practice, In: Proceedings of the First International Conference on Multiagent Systems, San Francisco (1995)
44. Rao, A.S. Georgeff, M.P.: Formal models and decision procedures for multi-agent systems, Technical report Technical Note 61, Australian AI Institute, 171 La Trobe Street, Melbourne, Australia (1995)
45. Weerasooriya, D. Rao, A. Ramamohanarao, K.: Design of a concurrent agent oriented language. In: Wooldridge M. Jennings N.R. (eds.) Intelligent Agents: Theories, Architectures, and Languages (LNAI), vol. 890, pp. 386–402. Springer, Heidelberg, Germany (1995)
46. Goodrich, K.H. McManus, J.W.: Development of A Tactical Guidance Research and Evaluation System(TGRES). AIAA Paper #89–3312, (August 1989)
47. McManus, J.W. Goodrich, K.H.: "Application of Artificial Intelligence (AI) Programming Techniques to Tactical Guidance for Fighter Aircraft." AIAA Paper #89–3525, (August 1989)
48. Goodrich, K.H. McManus J.W.: "An Integrated Environment For Tactical Guidance Research and Evaluation." AIAA Paper #90–1287 (May 1990)
49. McManus, J.W. : "A Parallel Distributed System for Aircraft Tactical Decision Generation." In: Proceedings of the Digital Avionics Systems Conference, Virginia Beach, VA (October 1990)
50. Ni, H.: Penny : BlackBoard Systems: The Blackboard Model of Problem solving and the Evolution Of Blackboard Architectures. AI Mag. 7(3), 38–53 (1986)
51. Burgin, G.H. et al.: An Adaptive Maneuvering Logic Computer Program for the Simulation of One-on-One Air-to-Air Combat. (NASA) vol I–II, pp. CR-2582–CR-2583 (1975)
52. Michael, W.: Intelligent agents. In: Weiss, G. (ed) Multiagent Systems: A Modern Approach to Distributed Artificial Intelligence, pp. 27–78. The MIT Press, Cambridge, USA (1999)
53. Lucas, A., Goss, S., The Potential For Intelligent Software Agents in defence simulation. In: Proceedings of the Information, Decision and Control, pp. 579–583, Adelaide, SA (1999)
54. Norling, E.J. Modelling Human Behaviour with BDI Agents, PhD. Thesis, Department of Computer Science and Engineering, University of Melbourne, (June 2009)
55. Boril, J. Jalovecky, R.: Experimental Identification of Pilot Response Using Measured Data from a Flight Simulator In: Proceedings of the 13th EANN (Engineering Applications of Neural Networks), LNCS AICT vol. 363(1), Springer, Berlin (2012)
56. Boril, J. Jalovecky, R.: Response of the Mechatronic System, Pilot—Aircraft on Incurred Step Disturbance. In: 53rd International Symposium ELMAR-2011, pp. 261–264. ITG, Zagreb (2011)
57. Jalovecky, R. Janu, P.: Human—Pilot's Features During Aircraft Flight Control from Automatic Regulation Viewpoint. In: 4th International Symposium on Measurement, Analysis and Modeling of Human Functions, pp. 119–123. Czech Republic: Czech Technical University in Prague, Prague (2010)
58. Jalovecky, R.: Man in the Aircraft's Flight Control System. Adv. Mil. Technol.—J. Sci., vol. 4(1), pp. 49–57 (2009)
59. Cameron, N., Thomson, D.G., Murray-Smith, D.J.: Pilot Modelling and Inverse Simulation for Initial Handling Qualities Assessment. Aeronaut.J. 107(1744), 511–520 (2003)
60. Boril, J. Jalovecky, R.: Simulation of Mechatronic System Pilot—Aircraft—Oscillation Damper. In: ICMT11—International Conference on Military Technologies, pp. 591–597. University of Defence, Brno (2011)

Chapter 6
Application of Gaussian Processes to the Modelling and Control in Process Engineering

Juš Kocijan and Alexandra Grancharova

Abstract Many engineering systems can be characterized as complex since they have a nonlinear behaviour incorporating a stochastic uncertainty. It has been shown that one of the most appropriate methods for modelling of such systems is based on the application of Gaussian processes (GPs). The GP models provide a probabilistic non-parametric modelling approach for black-box identification of nonlinear stochastic systems. This chapter reviews the methods for modelling and control of complex stochastic systems based on GP models. The GP-based modelling method is applied in a process engineering case study, which represents the dynamic modelling and control of a laboratory gas–liquid separator. The variables to be controlled are the pressure and the liquid level in the separator and the manipulated variables are the apertures of the valves for the gas flow and the liquid flow. GP models with different regressors and different covariance functions are obtained and evaluated. A selected GP model of the gas–liquid separator is further used to design an explicit stochastic model predictive controller to ensure the optimal control of the separator.

J. Kocijan
Department of Systems and Control, Jožef Stefan Institute, Jamova cesta 39
1000 Ljubljana, Slovenia
e-mail: jus.kocijan@ijs.si

J. Kocijan
Centre for Systems and Information Technologies, University of Nova Gorica,
Vipavska 13, 5000 Nova Gorica, Slovenia

A. Grancharova (✉)
Institute of System Engineering and Robotics, Bulgarian Academy of Sciences,
Acad. G. Bonchev Street, Bl. 2, P. O. Box 79, 1113 Sofia, Bulgaria
e-mail: alexandra.grancharova@abv.bg

V. E. Balas et al. (eds.), *Innovations in Intelligent Machines-5*,
Studies in Computational Intelligence 561, DOI: 10.1007/978-3-662-43370-6_6,
© Springer-Verlag Berlin Heidelberg 2014

6.1 Introduction

The technological development is following exponential growth and on one hand increasingly complex systems appear while on the other hand increasingly complex systems can be handled with new technologies. Number of engineering systems can be characterized as complex since they have a dynamic and nonlinear behaviour incorporating a stochastic uncertainty. Dynamic-systems control design utilises various kinds of computational intelligence methods for model development that result in so-called black-box models. Gaussian-process (GP) models provide a probabilistic, nonparametric modelling approach for black-box identification of nonlinear dynamic systems. They can highlight areas of the input space where model-prediction quality is poor, due to the lack of data or its complexity, by indicating the higher variance around the predicted mean. This property can be incorporated in the closed-loop control design. Gaussian-process models contain noticeably less coefficients to be optimised than parametric models that are frequently used in control design. This modelling method is not suggested as a replacement to any existing systems identification method, but rather as a complementary approach to modelling.

The aim of this chapter is to review some of the methods for modelling and control of complex stochastic systems based on GP models and to demonstrate a control based on Gaussian process model.

The chapter is structured as follows. First the modelling with Gaussian processes in general and the modelling of dynamic systems with Gaussian process models is explained. Then a review of inverse dynamics control, model-based predictive control and adaptive control based on Gaussian process models is given. The control design is illustrated with an example showing the design of explicit stochastic model-based predictive control of the gas–liquid separator. This example demonstrates developing of the model of plant with two signal inputs and two signal outputs, design of the controller for reference tracking and the closed-loop performance analysis with a computer simulation study.

6.2 Systems Modelling with Gaussian Processes

A GP model is a flexible, probabilistic, nonparametric model for the prediction of output-variable distributions. Its properties and application potentials are reviewed in [59].

A Gaussian process is a collection of random variables that have a joint multivariate Gaussian distribution. Assuming a relationship of the form $y = f(\mathbf{z})$ between the input $\mathbf{z}$ and the output y, we have $y_1, \ldots, y_N \sim \mathcal{N}(0, \Sigma)$, where $\Sigma_{pq} = \text{Cov}(y_p, y_q) = C(\mathbf{z}_p, \mathbf{z}_q)$ gives the covariance between the output points corresponding to the input points $\mathbf{z}_p$ and $\mathbf{z}_q$. Thus, the mean $\mu(\mathbf{z})$ and the covariance function $C(\mathbf{z}_p, \mathbf{z}_q)$ fully specify the Gaussian process.

The value of the covariance function $C(\mathbf{z}_p, \mathbf{z}_q)$ expresses the correlation between the individual outputs $f(\mathbf{z}_p)$ and $f(\mathbf{z}_q)$ with respect to the inputs $\mathbf{z}_p$ and $\mathbf{z}_q$. Note that the covariance function $C(\cdot\,,\,\cdot)$ can be any function that generates a positive semi-definite covariance matrix. It is usually composed of two parts,

$$C(\mathbf{z}_p, \mathbf{z}_q) = C_f(\mathbf{z}_p, \mathbf{z}_q) + C_n(\mathbf{z}_p, \mathbf{z}_q), \tag{6.1}$$

where C_f represents the functional part and describes the unknown system we are modelling, and C_n represents the noise part and describes the model of the noise.

For the noise part it is most common to use the constant covariance function, presuming white noise. The choice of the covariance function for the functional part also depends on the stationarity of the process. Assuming stationary data the most commonly used covariance function is the square exponential covariance function. The composite covariance function is therefore

$$C(\mathbf{z}_p, \mathbf{z}_q) = v_1 \exp\left[-\frac{1}{2}\sum_{d=1}^{D} w_d (z_{dp} - z_{dq})^2\right] + \delta_{pq} v_0, \tag{6.2}$$

where w_d, v_1 and v_0 are the 'hyperparameters' of the covariance function, D is the input dimension, and $\delta_{pq} = 1$ if $p = q$ and 0 otherwise. In contrast, assuming non-stationary data the polynomial or its special case, the linear covariance function, can be used. Other forms and combinations of covariance functions suitable for various applications can be found in [59]. The hyperparameters can be written as a vector $\Theta = [w_1, \ldots, w_D, v_1, v_0]^T$. The parameters w_d indicate the importance of the individual inputs: if w_d is zero or near zero, it means the inputs in dimension d contain little information and could possibly be neglected.

To accurately reflect the correlations present in the training data, the hyper-parameters of the covariance function need to be optimised. Due to the probabilistic nature of the GP models, the common model optimisation approach, where model parameters and possibly also the model structure are optimised through the minimization of a cost function defined in terms of model error (e.g., mean square error), is not readily applicable. A probabilistic approach to the optimisation of the model is more appropriate. Actually, instead of minimizing the model error, the probability of the model is maximised.

GP models can be easily utilized for a regression calculation. Consider a matrix $\mathbf{Z}$ of N D-dimensional input vectors where $\mathbf{Z} = [\mathbf{z}_1, \mathbf{z}_2, \ldots, \mathbf{z}_N]^T$ and a vector of the output data $\mathbf{y} = [y_1, y_2, \ldots, y_N]$. Based on the data $(\mathbf{Z}, \mathbf{y})$, and given a new input vector $\mathbf{z}^*$, we wish to find the predictive distribution of the corresponding output y^*. Based on the training set $\mathbf{Z}$, a covariance matrix $\mathbf{K}$ of size $N \times N$ is determined. The overall problem of learning unknown parameters from data corresponds to the predictive distribution $p(y^*|\mathbf{y}, \mathbf{Z}, \mathbf{z}^*)$ of the new target y, given the training data $(\mathbf{y}, \mathbf{Z})$ and a new input $\mathbf{z}^*$. In order to calculate this posterior distribution, a prior distribution over the hyperparameters $p(\Theta|\mathbf{y}, \mathbf{Z})$ can first be defined, followed by the integration of the model over the hyperparameters

$$p(y^*|\mathbf{y}, \mathbf{Z}, \mathbf{z}^*) = \int p(y^*|\Theta, \mathbf{y}, \mathbf{Z}, \mathbf{z}^*)p(\Theta|\mathbf{y}, \mathbf{Z})d\Theta. \qquad (6.3)$$

The computation of such integrals can be difficult due to the intractable nature of the non-linear functions. A solution to the problem of intractable integrals is to adopt numerical integration methods such as the Monte-Carlo approach. Unfortunately, significant computational efforts may be required to achieve a sufficiently accurate approximation.

In addition to the Monte-Carlo approach, another standard and general practice for estimating hyperparameters is the maximum-likelihood estimation, i.e., to minimize the following negative log-likelihood function [59]:

$$\mathscr{L}(\Theta) = -\frac{1}{2}\log(|\mathbf{K}|) - \frac{1}{2}\mathbf{y}^T\mathbf{K}^{-1}\mathbf{y} - \frac{N}{2}\log(2\pi) \qquad (6.4)$$

As the likelihood is, in general, non-linear and multi-modal, efficient optimisation routines usually entail the gradient information. The computation of the derivative of $\mathscr{L}$ with respect to each of the parameters is as follows

$$\frac{\partial\mathscr{L}(\Theta)}{\partial\theta_i} = -\frac{1}{2}\text{trace}\left(\mathbf{K}^{-1}\frac{\partial\mathbf{K}}{\partial\theta_i}\right) + \frac{1}{2}\mathbf{y}^T\mathbf{K}^{-1}\frac{\partial\mathbf{K}}{\partial\theta_i}\mathbf{K}^{-1}\mathbf{y}. \qquad (6.5)$$

For performing a regression, the availability of the training set $\mathbf{Z}$ and the corresponding output set $\mathbf{y}$ is assumed. Based on the training set $\mathbf{Z}$, a covariance matrix $\mathbf{K}$ of size $N \times N$ is determined. The aim is to find the distribution of the corresponding output y^* for some new input vector $\mathbf{z}^* = [z_1(N+1), z_2(N+1), \ldots, z_D(N+1)]^T$.

For the collection of random variables $[y_1, \ldots, y_N, y^*]$ we can write:

$$[\mathbf{y}, y^*] \sim \mathscr{N}(0, \mathbf{K}^*) \qquad (6.6)$$

with the covariance matrix

$$\mathbf{K}^* = \begin{bmatrix} \mathbf{K} & \mathbf{k}(\mathbf{z}^*) \\ \mathbf{k}^T(\mathbf{z}^*) & \kappa(\mathbf{z}^*) \end{bmatrix} \qquad (6.7)$$

where $\mathbf{y} = [y_1, \ldots, y_N]$ is a $1 \times N$ vector of training targets. The predictive distribution of the output for a new test input has a normal probability distribution with a mean and variance

$$\mu(y^*) = \mathbf{k}(\mathbf{z}^*)^T\mathbf{K}^{-1}\mathbf{y}, \qquad (6.8)$$

$$\sigma^2(y^*) = \kappa(\mathbf{z}^*) - \mathbf{k}(\mathbf{z}^*)^T\mathbf{K}^{-1}\mathbf{k}(\mathbf{z}^*), \qquad (6.9)$$

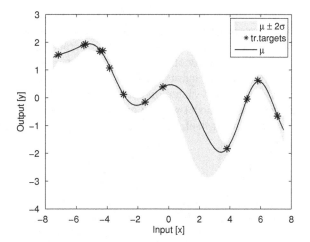

Fig. 6.1 Using GP models: in addition to the prediction mean value (*full line*), we obtain a 95 % confidence region (*gray band*) for the underlying function f

where $k(\mathbf{z}^*) = [C(\mathbf{z}_1, \mathbf{z}^*), ..., C(\mathbf{z}_N, \mathbf{z}^*)]^T$ is the $N \times 1$ vector of covariances between the test and training cases, and $\kappa(z^*) = C(\mathbf{z}^*, \mathbf{z}^*)$ is the covariance between the test input itself.

The obtained model, in addition to the mean value, provides information about the confidence in the prediction by the variance. Usually, the confidence of the prediction is depicted with a 2σ interval, which is an about 95 % confidence interval. This confidence region can be seen in the example in Fig. 6.1 as a grey band. It highlights the areas of the input space where the prediction quality is poor, due to the lack of data or noisy data, by indicating a wider confidence band around the predicted mean.

The cross-validation response fit is usually evaluated by performance measures. Beside commonly used performance measures such as *e.g.* mean relative square error:

$$\text{MRSE} = \sqrt{\frac{\sum_{i=1}^{N} e_i^2}{\sum_{i=1}^{N} y_i^2}}, \tag{6.10}$$

where y_i and $e_i = \hat{y}_i - y_i$ are the ith system's output and prediction error, the performance measures such as log predictive density error [2, 35, 59]:

$$\text{LPD} = \frac{1}{2}\log(2\pi) + \frac{1}{2N}\sum_{i=1}^{N}(\log(\sigma_i^2) + \frac{e_i^2}{\sigma_i^2}), \tag{6.11}$$

where σ_i^2 is the ith prediction variance, can be used for evaluating GP models. It takes into account not only mean prediction but the entire predicted distribution. Another possible performance measure is the negative log-likelihood $\mathscr{L}$ of the training data [cf. (6.4)]. $\mathscr{L}$ is the measure inherent to the hyperparameter optimisation process and gives the likelihood that the training data is generated by given, i.e. trained, model. Therefore it is applicable for validation on identification data only. The smaller the MRSE, LPD and $\mathscr{L}$ are, the better the model is.

GP models can, like neural networks, be used to model static non-linearities and can therefore be used for the modelling of dynamic systems [2, 35, 38] as well as time series, if lagged samples of the output signals are fed back and used as regressors. A review of recent developments in the modelling of dynamic systems using GP models and its applications can be found in [34].

A single-input single-output dynamic GP model is trained as the nonlinear autoregressive model with an exogenous input (NARX) representation, where the output at time instant k depends on the delayed output y and the exogenous control input u:

$$y(k) = f_S(y(k-1), \ldots, y(k-n), u(k-1), \ldots, u(k-n)) + \xi(k) \qquad (6.12)$$

where f_S denotes a function, $\xi(k)$ is white noise disturbance with normal distribution and the output $y(k)$ depends on the state vector $\mathbf{x}(k) = [y(k-1), y(k-2), \ldots, y(k-n), u(k-1), u(k-2), \ldots, u(k-n)]$ at time instant k. This model notation can be generalised to mutivariable cases, i.e., cases with multiple inputs and outputs. Inputs and outputs at time instant k are in the multivariable cases represented with vectors of values.

$$\mathbf{y}(k) = f(\mathbf{y}(k-1), \ldots, \mathbf{y}(k-n), \mathbf{u}(k-1), \ldots, \mathbf{u}(k-n)) + \boldsymbol{\xi}(k) \qquad (6.13)$$

where f denotes a function and $\boldsymbol{\xi}(k)$ are Gaussian random variables.

For the validation of obtained dynamic GP model the nonlinear output-error (NOE), also called parallel, model is used. This means that the NARX model is used to predict a further step ahead by replacing the data at instant k with the data at instant $k + 1$ and using the prediction $\hat{y}(k)$ from the previous prediction step instead of the measured $y(k)$. This is then repeated indefinitely. The latter possibility is equivalent to simulation. Simulation, therefore, means that only on the basis of previous samples of a process input signal $u(k - i)$ can the model simulate future outputs. Frequently, the mean value of prediction $\hat{y}(k)$ is used to replace $y(k)$, which is called 'naive' simulation. Other possibilities, where the entire distribution is used, are described in, e.g., [38].

6.3 Control Algorithms Based on Gaussian Process Models

This section provides a review of some of the control methods that are based on GP model. Reader is referred to [33] for more comprehensive review.

6.3.1 Inverse Dynamics Control

When machine learning methods are preliminary introduced for control, frequently the following scheme appears. An inverse model of the process is developed to be

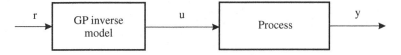

Fig. 6.2 General block scheme of the direct inverse control

connected in series with the process and therefore an open-loop control system is formed. This kind of approach is usually not meant as effective control solution, but mainly as a demonstration of particular machine learning method.

The basic principle in brief is as follows. If the system to be controlled can be described by input-output model

$$\mathbf{y}(k+1) = f(\mathbf{y}(k), \ldots, \mathbf{y}(k-n+1), \ldots, \mathbf{u}(k), \ldots, \mathbf{u}(k-m)) \qquad (6.14)$$

then the corresponding inverse model is

$$\hat{\mathbf{u}}(k) = \hat{f}^{-1}(\mathbf{y}(k+1), \mathbf{y}(k), \ldots, \mathbf{y}(k-n+1), \ldots, \mathbf{u}(k-1), \ldots, \mathbf{u}(k-m)) \quad (6.15)$$

where the notation $\hat{\cdot}$ denotes the estimator.

Assuming that this inverse system has been obtained it can be used to generate control input that approaches desired process output, when the reference input is given to the inverse model. This means that samples of $\mathbf{y}$ in Eq. (6.15) are replaced by reference values $\mathbf{r}$.

The principle is illustrated in Fig. 6.2.

There is a list of assumptions and constraints that need to be satisfied for a such system to be practical implemented. The assumptions necessary for the open-loop control to be operational are: no disturbances in the system, no uncertainties and changes in the process and open-loop controller that is the perfect inverse of the process in the region of operation. Since these assumptions are frequently not fulfilled in the real world, the inverse system is usually realised as the adaptive system, where the controller matches any changes in the process on-line.

Training of the inverse model requires that the process and inverse model are input-output stable. This is because signals are always constrained in magnitude, which disables the open-loop control of unstable systems. Even in the case of computer simulation, inputs and outputs can not be infinitely large.

When the mentioned assumptions are satisfied, the inverse model can be modelled from appropriately selected outputs and inputs of the process following Eq. (6.15). In the case of open-loop controller realisation with GP model, only the mean value of controller output prediction is the input into the process to be controlled.

Most publications on inverse dynamics control based in GP models describe dynamic systems of single-input single-output type. The reinforcement learning of the described open-loop controller actions based on GP model is introduced in [14]. The entire system is implemented only as a computer simulation and meant to demonstrate reinforcement learning, rather then practically applicable control system principle.

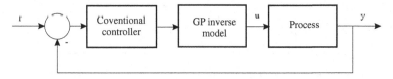

Fig. 6.3 General block scheme of the inverse dynamics control

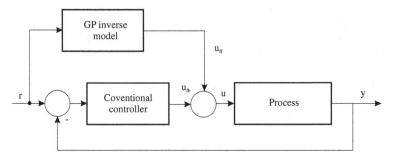

Fig. 6.4 General block scheme of existing control system with GP inverse model as feedforward for improvement of closed-loop performance

Different approach to open-loop control is given in [32] where GP model in the role of open-loop controller is taught with reinforcement learning to mimic the outputs of an optimal and closed-loop conventional controller.

Another method that uses inverse model for cancelling nonlinearities of the process to be controlled is Inverse Dynamics Control [51] illustrated in Fig. 6.3.

This is the closed-loop method that contains conventional controller to deal with miss-matches between nonlinearity compensator and process as well as with the process disturbances. Such scheme with different sorts of inverse models is commonly used for dynamics control in robotics. The inverse model can be identified off-line or on-line. The application with the GP model of inverse process dynamics that is identified off-line is given in [51] for a robot control investigation.

Feed forward that eliminates the process nonlinearities is another control method that is used mainly in robotics. Its principle is depicted in Fig. 6.4.

The control signal consists of feedforward and feedback component $u = u_{ff} + u_{fb}$. The feedback loop with a conventional, frequently linear, controller is required to maintain stability and disturbance rejection for this control system purposed for set-point tracking. The feedforward that is a stable inverse model is meant to compensate for process nonlinearities. The inverse model has to be as precise as possible in the region of operation to enable the required performance. The closed-loop performance is deteriorated in the case of unmodelled nonlinearities. The feedforward is generally considered as a function of the desired set-point, in the case of robotic control that would mean desired robot trajectories.

The concept has some practical advantages [53]. First, the data necessary for the inverse model can be collected from beforehand assembled closed-loop system without the feedforward component. Second, the feedforward signal can be introduced gradually during control system implementation from the reason of cautiousness. Third, in the case that the inverse dynamics is to be avoided, only static feedforward can be used with feedback controller compensating for erroneous feedforward signal.

The inverse model can be identified off-line or on-line. The case when inverse GP model is identified off line and used in such control set-up is described in [50] and [51]. The adaptive cases are mentioned in Sect. 6.3.3.

None of these applications of inverse GP models uses entire information from prediction distributions rather they focus on the predictions' mean values. In such a way the full potential of GP models for this kind of control is not utilised entirely, e.g., information of variances could be used for maintaining or indicating the region of the nominal closed-loop performance.

6.3.2 Model-Based Predictive Control

Model Predictive Control (MPC) [45] is a common name for computer control algorithms that use an explicit process model to predict the future plant response. According to this prediction in the chosen period, also known as the prediction horizon, the MPC optimises the future plant behaviour by solving a finite-horizon optimal control problem. Since the state of the system is updated during each sampling period, a new optimization problem must be solved at each sampling interval. This is known as the receding horizon strategy (RHC). The popularity of MPC is to a great extent owed to the ability of MPC algorithms to deal both with state and input constraints that are frequently met in control practice and are often not well addressed with other approaches.

In our case we are interested in the applications of Nonlinear Model Predictive Control (NMPC) principle with a GP model. Stochastic NMPC problems are formulated in the applications where the system to be controlled is described by a stochastic model such as the GP model. Stochastic problems like state estimation are studied for long time, but, in our case, we explore only stochastic NMPC problem. Nevertheless, most known stochastic MPC approaches are based on parametric probabilistic models. Alternatively, the stochastic systems can be modeled with nonparametric models which can offer a significant advantage compared to the *parametric* models. This is related to the fact that the *nonparametric* probabilistic models, like Gaussian process models, provide information about prediction uncertainties which are difficult to evaluate appropriately with the parametric models.

The nonlinear model predictive control as it is applied with the GP model can be in general described with a block diagram, as depicted in Fig. 6.5. The model is fixed, identified off-line, which means that the resulting control algorithm is not an adaptive one. The structure of the entire control loop is therefore less complex as in the case of adaptive control.

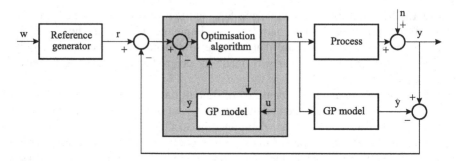

Fig. 6.5 Block diagram of model predictive control system

The control objective is to be achieved by minimization of the cost function. The cost function penalises deviations of the predicted controlled outputs $\hat{\mathbf{y}}(k+j|k)$ from a vector of reference trajectories $\mathbf{r}(k+j|k)$. These reference trajectories may depend on measurements made up to time instant k. Its initial point may be the vector of measurements at outputs $\mathbf{y}(k)$, but can also be a fixed set-point, or some predetermined trajectory. The minimization of cost function, in which the matrix containing vectors of future control signals ($\mathbf{U}$) is calculated, can be subject to various constraints (e.g., input, state, rates, etc.).

MPC solves a constrained control problem. The multiple-input, multiple-output case is elaborated here. A stochastic nonlinear discrete-time system can be described in the input-output form or in the very common state-space form:

$$\mathbf{x}(k+1) = g(\mathbf{x}(k), \mathbf{u}(k)) + \xi_1(k) \qquad (6.16)$$

$$\mathbf{y}(k) = h(\mathbf{x}(k), \mathbf{u}(k)) + \xi_2(k) \qquad (6.17)$$

where $\mathbf{x} \in \mathbb{R}^n$, $\mathbf{u} \in \mathbb{R}^m$ and $\mathbf{y} \in \mathbb{R}^m$ are the state, input and output variables respectively, $\xi_i(k)$; $i = 1, 2$ are Gaussian random variables representing disturbances, and g, h are nonlinear continuous functions.

Corresponding input and state constraints of the general form are:

$$\mathbf{u}(k) \in U \qquad (6.18)$$

$$\mathbf{x}(k) \in X \qquad (6.19)$$

where U and X are sets of admissible inputs and feasible states respectively. Then, the optimisation problem is:

$$V^*(k) = \min_{\mathbf{U}} J(\mathbf{U}, \mathbf{x}(k), \mathbf{r}(k), \mathbf{u}(k-1)) \qquad (6.20)$$

with $\mathbf{U} = [\mathbf{u}(k), \mathbf{u}(k+1), ..., \mathbf{u}(k+N-1)]$ and the cost function given by the following general form:

$$J(\mathbf{U}, \mathbf{x}(k), \mathbf{r}(k), \mathbf{u}(k-1)) = \sum_{j=0}^{N-1} L(\hat{\mathbf{x}}(k+j|k), \mathbf{u}(k+j)) \qquad (6.21)$$

where $\hat{\mathbf{x}}(k+j|k)$ is the predicted state, L is a nonlinear continuous function and it is assumed that the cost falls to zero once the state has entered the set of optimal states X_o, namely $L(\mathbf{x}, \mathbf{U}) = 0$ if $\mathbf{x} \in X_0$. The following terminal constraint is imposed:

$$\hat{\mathbf{x}}(k+N|k) \in X_0. \qquad (6.22)$$

This is a general form and MPC formulations vary with various models, cost functions and parameters.

Frequently used cost function in MPC literature is:

$$J(k) = ||\mathrm{E}\{\hat{\mathbf{y}}(k+N|k)\} - \mathbf{r}(k)||_{\mathbf{P}}^2 +$$
$$\sum_{j=0}^{N-1} \left[||\mathrm{E}\{\hat{\mathbf{y}}(k+j|k)\} - \mathbf{r}(k)||_{\mathbf{Q}}^2 + ||\Delta\mathbf{u}_{k+j}||_{\mathbf{R}}^2 \right] \qquad (6.23)$$

where N is a finite horizon and $||\mathbf{x}||_{\mathbf{A}} = \sqrt{\mathbf{x}^{\mathrm{T}}\mathbf{A}\mathbf{x}}$; $\mathbf{A} = \mathbf{P}, \mathbf{Q}, \mathbf{R}$ are positive definite matrices and the notation $\mathrm{E}\{\cdot\}$ denotes an 'expectation' conditional upon data available up to and including current time instant k.

There are many alternative ways of how NMPC can be realised with Gaussian process models:

Cost function. The cost function (6.21) is a general one and various special cost functions can be derived out of it. It is well known that the selection of the cost function has a major impact on the amount of computation.

Optimization problem for $\Delta\mathbf{u}$ instead of $\mathbf{u}$. This is not just a change of the formalism, but also enables forms of MPC containing integral action.

Process model. The process model can be determined off-line and fixed for the time of operation or determined on-line during the operation of controller. The on-line model identification is described in Sect. 6.3.3.

Soft constraints. Using constraint optimization algorithms is very demanding for computation and soft constrains. In other words weights on constrained variables in cost function, can be used to decrease the amount of computation. More on this topic can be found in [45].

Linear MPC. It is worth to remark that even though this is a constrained nonlinear MPC problem it can be used in its specialised form as a robust linear MPC.

Various predictive control methods can be applied with GP models depending on designers choice and imposed constraints. Using GP models does not impose any particular constraint on cost function, optimisation method or any other element of choice for predictive control design.

An application of model predictive control with the GP model using the general cost function described with Eq. (6.23) can be found in [21]. Investigations of three special forms of MPC are more frequent in the literature. These three algorithms are: Internal Model Control (IMC), Predictive Functional Control (PFC), and the approximate explicit control, which are described in subsequent subsections. A principle different from the listed, where control is based on the estimation and the multiple-step prediction of system output in combination with fuzzy models is given in [54].

6.3.2.1 Internal Model Control

In this strategy the controller is chosen to be an inverse of the plant model. Internal model control is one of the most commonly used model–based techniques for the control of nonlinear systems. It can be considered also as the simplest form of MPC with prediction and control horizon equal to one step. IMC with the GP model is elaborated in [25, 26, 28] and [29]. The description of IMC with the GP model hereafter is adopted from these references.

The main difference between the various internal model control approaches is in the choice of the internal model and its inverse. It was shown in [27] that the GP model based on the squared-exponential covariance function is not analytically invertible. Instead of calculating the exact inverse, a numerical approach such as successive approximation or Newton-Raphson optimisation method, can be used to find the control effort to solve the following equation:

$$f(\mathbf{y}(k), \ldots, \mathbf{y}(k-n), \mathbf{u}(k), \ldots, \mathbf{u}(k-m)) - \mathbf{q}(k) = 0, \qquad (6.24)$$

where

$$f(\mathbf{y}(k), \ldots, \mathbf{y}(k-n), \mathbf{u}(k), \ldots, \mathbf{u}(k-m)) = \hat{\mathbf{y}}(k+1) \qquad (6.25)$$

and $\mathbf{q}(k)$ is the controller input.

The GP model is trained as a one-step ahead prediction model. This GP model is then included in the IMC structure and the numerical inverse of the Eq. (6.24), considering only mean values of model predictions, is found at each sample. The IMC works well when the control input and the output of the system are in the region where the model was trained. As soon as the system moves away from the well modelled region this can cause sluggish and, in certain cases, also unstable closed-loop system behaviour.

Since poor closed-loop performance is the result of the model being driven outside its trained region, the naive approach would be to constrain the control input. When the system is driven in this untrained portion of the operating space, the increase of the predicted variance will indicate a reduced confidence in the prediction. This increase of variance can be used as a constraint in the optimisation algorithm utilised to solve Eq. (6.24). This concept is shown in Fig. 6.6.

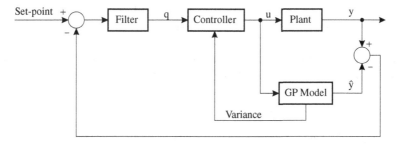

Fig. 6.6 Variance-constrained internal model control structure

The basic idea of the algorithm is to optimise the control effort so that the variance does not increase above its predefined limit. Since the GP model is not analytically invertible and numerical approaches have to be utilised to find the inverse of the model at each sample time, the associated computation load rises rapidly with the number of training data points. This is the main drawback of the GP modelling approach for IMC.

6.3.2.2 Predictive Functional Control

Predictive functional control is a form of the predictive control that in principle is no different to a general predictive control. Its distinct features are relatively low number of so called coincidence points, the use of a reference trajectory, which is distinct from the set-point trajectory and the assumption that the future input is a linear combination of a few simple basis functions. More details can be found in, e.g., [45].

In the following description the PFC with one coincidence point and constant output within the control horizon is used. Variants of this kind of predictive control with the GP model are described in [37, 39, 40, 41, 44] and [3]. The predictive control based on the GP model was for the first time introduced in published reference in [41].

A moving-horizon minimisation problem of the form [45]

$$J_{\text{opt}}(k) = \min_{\mathbf{U}} \|\mathbf{r}(k+P) - \mu(\hat{\mathbf{y}}(k+P|k))\|^2 \qquad (6.26)$$

subject to constraints on the output variance, the input hard constraints, the input rate constraints, the state hard constraints and the state rate constraints, is applied as the first presented choice, where $\mathbf{U} = [\mathbf{u}(k), \ldots, \mathbf{u}(k+P)]$ is the matrix of input signal up to the coincidence point P, which is the point where a match between output and the reference value is expected, $\mu(\hat{\mathbf{y}}(k+P|k)) = [\mathrm{E}\{\hat{y}_1(k+P|k)\}, \ldots,$ $\mathrm{E}\{\hat{y}_m(k+P|k)\}]^{\mathrm{T}}$ and $\|\mathbf{x}\| = \sqrt{\mathbf{x}^{\mathrm{T}}\mathbf{x}}$. The listed constraints are in general functions of some scheduling variable in the general form, but are many times set to be constant values. The process model is a GP model.

The optimisation algorithm, which is constrained nonlinear programming, is solved at each sample time over a prediction horizon of length P, for a series of moves which equals to control horizon.

A possible alternative selection of the cost function [39] that avoids constrained optimisation and is therefore computationally less demanding would be

$$J(k) = \mathrm{E}\{\|\mathbf{r}(k+P) - \hat{\mathbf{y}}(k+P)\|^2\}. \qquad (6.27)$$

where output variances become part of the cost function. The control strategy with cost function (6.27) is 'to avoid' going into regions with higher variance. The term 'higher variance' does not specify any specific value. In the case that controller does not seem to be 'cautious' enough, a 'quick-and-dirty' option is that the variance term can be weighted to enable shaping of the closed-loop response according to variance information. Beside the difference in the optimisation algorithm the presented options give also a design choice on how 'safe' the control algorithm is. In the case when it is very undesirable to go into 'unknown' regions the constrained version may be better option.

6.3.2.3 Approximate Explicit Stochastic Nonlinear Model Predictive Control

The MPC formulation described up-to-now provides the control action $\mathbf{u}(k)$ as a function of states $\mathbf{x}(k)$ defined *implicitly* by the cost function and constraints. In the last decade, several methods for *explicit* solution of MPC problems have been suggested (see for example [1, 19, 56]). The main motivation behind *explicit* MPC is that an *explicit* state feedback law avoids the need for executing a numerical optimization algorithm in real time, and is therefore potentially useful for applications where MPC has not traditionally been used. By treating $\mathbf{x}(k)$ as a vector of parameters, the goal of the explicit methods is to solve the MPC problem off-line with respect to all values of $\mathbf{x}(k)$ of interest and make the dependence of the control input on the state explicit. It has been shown in [4] that the feedback solution to MPC problems for constrained linear systems has an explicit representation as a piecewise linear (PWL) state feedback defined on a polyhedral partition of the state space. The benefits of an explicit solution, in addition to the efficient on-line computations, include also verifiability of the implementation and the possibility to design embedded control systems with low software and hardware complexity. For the nonlinear and stochastic MPC the benefits of explicit solutions are even higher than for linear MPC, since the computational efficiency and verifiability are even more important. In [19], approaches for off-line computation of explicit sub-optimal piecewise predictive controllers for general nonlinear systems with state and input constraints have been presented, based on the multi-parametric Non-linear Programming (mp-NLP) ideas [15].

In [23], an approximate mp-NLP approach to off-line computation of explicit suboptimal stochastic NMPC controller for constrained nonlinear systems based on a GP model (abbreviated as GP-NMPC) has been proposed. The approach represents an extension of the approximate methods in [20] and [31]. The approximate explicit GP-NMPC approach has been elaborated in [24].

Formulation of the GP-NMPC Problem as an mp-NLP Problem

Consider a stochastic nonlinear discrete-time system (6.16):

$$\mathbf{x}(k+1) = g(\mathbf{x}(k), \mathbf{u}(k)) + \xi(k)$$

where $\mathbf{x}(k) \in \mathbb{R}^n$ and $\mathbf{u}(k) \in \mathbb{R}^m$ are the state and input variables, $\xi(k) \in \mathbb{R}^n$ are Gaussian disturbances, and $g : \mathbb{R}^n \times \mathbb{R}^m \to \mathbb{R}^n$ is a nonlinear continuous function. Suppose that a Gaussian process model of the system (6.16) is obtained by applying the approach described in Sect. 6.2. Suppose the initial state $\mathbf{x}(k) = \mathbf{x}(k|k)$ and the control inputs $\mathbf{u}(k+j)$, $j = 0, 1, \ldots, N-1$ are given. Then, the probability distribution of the predicted states $\mathbf{x}(k+j+1|k)$, $j = 0, 1, \ldots, N-1$ which correspond to the given initial state $\mathbf{x}(k|k)$ and control inputs $\mathbf{u}(k+j)$, $j = 0, 1, \ldots, N-1$ can be obtained [17]:

$$\begin{aligned} \mathbf{x}(k+j+1|k)|\mathbf{x}(k+j|k), \ \mathbf{u}(k+j) \\ \sim \mathcal{N}(\mu(\mathbf{x}(k+j+1|k)), \sigma^2(\mathbf{x}(k+j+1|k))), \ j = 0, 1, \ldots, N-1 \end{aligned} \tag{6.28}$$

The 95 % confidence interval of the random variable $\mathbf{x}(k+j+1|k)$ is $[\mu(\mathbf{x}(k+j+1|k)) - 2\sigma(\mathbf{x}(k+j+1|k));$ $\mu(\mathbf{x}(k+j+1|k)) + 2\sigma(\mathbf{x}(k+j+1|k))]$, where $\sigma(\mathbf{x}(k+j+1|k))$ is the standard deviation.

In [22], a *disturbance rejection* GP-NMPC problem is considered, where the goal is to steer the state vector $\mathbf{x}(k)$ to the origin. Suppose that a full measurement of the state $\mathbf{x}(k)$ is available at the current time k. For the current $\mathbf{x}(k)$, the disturbance rejection GP-NMPC solves the following optimization problem:

Problem 1

$$V^*(\mathbf{x}(k)) = \min_{\mathbf{U}} J(\mathbf{U}, \mathbf{x}(k)) \tag{6.29}$$

subject to $\mathbf{x}(k|k) = \mathbf{x}(k)$ and:

$$\mu(\mathbf{x}(k+j|k)) - 2\sigma(\mathbf{x}(k+j|k)) \geq \mathbf{x}_{\min}, j = 1, \ldots, N \tag{6.30}$$

$$\mu(\mathbf{x}(k+j|k)) + 2\sigma(\mathbf{x}(k+j|k)) \leq \mathbf{x}_{\max}, j = 1, \ldots, N \tag{6.31}$$

$$\mathbf{u}_{\min} \leq \mathbf{u}(k+j) \leq \mathbf{u}_{\max}, j = 0, 1, \ldots, N-1 \tag{6.32}$$

$$\max \{ \| \mu(\mathbf{x}(k \mid N|k)) - 2\sigma(\mathbf{x}(k+N|k)) \|, \\ \| \mu(\mathbf{x}(k+N|k)) + 2\sigma(\mathbf{x}(k+N|k)) \| \} \leq \delta \tag{6.33}$$

$$\mathbf{x}(k+j+1|k)|\mathbf{x}(k+j|k), \mathbf{u}(k+j) \sim \mathcal{N}(\mu(\mathbf{x}(k+j+1|k)), \sigma^2(\mathbf{x}(k+j+1|k)))$$
$$j = 0, 1, \ldots, N-1$$

$$\tag{6.34}$$

with $\mathbf{U} = [\mathbf{u}(k), \mathbf{u}(k+1), \ldots, \mathbf{u}(k+N-1)]$ and the cost function given by:

$$J(\mathbf{U}, \mathbf{x}(k)) = \sum_{j=0}^{N-1} \left[\| \mu(\mathbf{x}(k+j|k)) \|_{\mathbf{Q}}^2 + \| \mathbf{u}(k+j) \|_{\mathbf{R}}^2 \right] + \| \mu(\mathbf{x}(k+N|k)) \|_{\mathbf{P}}^2 \tag{6.35}$$

Here, N is a finite horizon and $\mathbf{P}, \mathbf{Q}, \mathbf{R} \succ 0$.

It should be noted that a more general stochastic MPC problem is formulated in [5–7, 42], where a probabilistic formulation of the cost includes the probabilistic bounds of the predicted variable. The stochastic MPC problem considered here (Problem 1) is of a more special form since the cost function (6.35) includes the mean value of the random variable. However, the approximate approach to the explicit solution of Problem 1 (which is based on the approximate mp-NLP algorithms, given in [19]) can be easily extended to the more general case of stochastic MPC problem formulation where the optimization is performed on the expected value of the cost function.

In [19, 23, 24], a *reference tracking* GP-NMPC problem is considered, where the goal is to have the state vector $\mathbf{x}(k)$ track the reference signal $\mathbf{r}(k) \in \mathbb{R}^n$. For the current $\mathbf{x}(k)$, the reference tracking GP-NMPC solves the following optimization problem:

Problem 2

$$V^*(\mathbf{x}(k), \mathbf{r}(k), \mathbf{u}(k-1)) = \min_{\mathbf{U}} J(\mathbf{U}, \mathbf{x}(k), \mathbf{r}(k), \mathbf{u}(k-1)) \tag{6.36}$$

subject to $\mathbf{x}(k|k) = \mathbf{x}(k)$ and:

$$\mu(\mathbf{x}(k+j|k)) - 2\sigma(\mathbf{x}(k+j|k)) \geq \mathbf{x}_{\min}, j = 1, \ldots, N \tag{6.37}$$

$$\mu(\mathbf{x}(k+j|k)) + 2\sigma(\mathbf{x}(k+j|k)) \leq \mathbf{x}_{\max}, j = 1, \ldots, N \tag{6.38}$$

$$\mathbf{u}_{\min} \leq \mathbf{u}(k+j) \leq \mathbf{u}_{\max}, j = 0, 1, \ldots, N-1 \tag{6.39}$$

$$\Delta\mathbf{u}_{\min} \leq \Delta\mathbf{u}(k+j) \leq \Delta\mathbf{u}_{\max}, j = 0, 1, \ldots, N-1 \tag{6.40}$$

$$\max \{\|\mu(\mathbf{x}(k+N|k)) - 2\sigma(\mathbf{x}(k+N|k)) - \mathbf{r}(k)\|$$
$$\|\mu(\mathbf{x}(k+N|k)) + 2\sigma(\mathbf{x}(k+N|k)) - \mathbf{r}(k)\|\} \leq \delta \tag{6.41}$$

$$\varDelta\mathbf{u}_{t+k} = \mathbf{u}(k+j) - \mathbf{u}(k+j-1), j = 0, 1, \ldots, N-1 \tag{6.42}$$

$$\mathbf{x}(k+j+1|k)|\mathbf{x}(k+j|k), \mathbf{u}(k+j) \sim \mathcal{N}(\mu(\mathbf{x}(k+j+1|k)),$$
$$\sigma^2(\mathbf{x}(k+j+1|k)))j = 0, 1, \ldots, N-1 \tag{6.43}$$

with $\mathbf{U} = [\mathbf{u}(k), \mathbf{u}(k+1), \ldots, \mathbf{u}(k+N-1)]$ and the cost function given by:

$$J(\mathbf{U}, \mathbf{x}(k), \mathbf{r}(k), \mathbf{u}(k-1)) = \sum_{j=0}^{N-1}\left[\|\mu(\mathbf{x}(k+j|k)) - \mathbf{r}(k)\|_{\mathbf{Q}}^2 + \|\varDelta\mathbf{u}(k+j)\|_{\mathbf{R}}^2\right]$$
$$+ \|\mu(\mathbf{x}(k+N|k)) - \mathbf{r}(k)\|_{\mathbf{P}}^2 \tag{6.44}$$

Similar to above, N is a finite horizon and $\mathbf{P}, \mathbf{Q}, \mathbf{R} \succ 0$ are weighting matrices. This formulation is also somewhat extended since it includes input-rate constraints and cost.

From a stability point of view it is desirable to choose δ in the terminal constraint (6.33) or (6.41) sufficiently small [46]. If the horizon N is large and the Gaussian process model has a small prediction uncertainty, then it is more likely that the choice of a small δ will be possible.

Using a direct single shooting strategy [43], the equality constraints are eliminated and the optimization Problems 1 and 2 can be formulated in a compact form as follows [19, 22, 24]:

Problem 3

$$V^*(\tilde{\mathbf{x}}) = \min_{\mathbf{U}} J(\mathbf{U}, \tilde{\mathbf{x}}) \text{ subject to } G(\mathbf{U}, \tilde{\mathbf{x}}) \leq 0 \tag{6.45}$$

Here $\tilde{\mathbf{x}}(k) \in \mathbb{R}^{\tilde{n}}$ is the parameter vector. For the regulation Problem 1 it is given by:

$$\tilde{\mathbf{x}}(k) = \tilde{\mathbf{x}}(k), \tilde{n} = n \tag{6.46}$$

while for the reference tracking Problem 2 it is:

$$\tilde{\mathbf{x}}(k) = [\mathbf{x}(k), \mathbf{r}(k), \mathbf{u}(k-1)] \in \mathbb{R}^{\tilde{n}}, \tilde{n} = 2n + m \tag{6.47}$$

Problem 3 defines an mp-NLP, since it is NLP in $\mathbf{U}$ parameterised by $\tilde{\mathbf{x}}$. An optimal solution to this problem is denoted $\mathbf{U}^* = [\mathbf{u}(k)^*, \mathbf{u}(k+1)^*, \ldots, \mathbf{u}(k+N-1)^*]$ and the control input is chosen according to the receding horizon policy $\mathbf{u}(k) = \mathbf{u}(k)^*$. Define the set of feasible parameter vectors as follows:

$$X_f = \{\tilde{\mathbf{x}} \in \mathbb{R}^{\tilde{n}} \,|\, G(\mathbf{U}, \tilde{\mathbf{x}}) \leq 0 \text{ for some } \mathbf{U} \in \mathbb{R}^{Nm}\} \qquad (6.48)$$

For Problem 1, X_f is the set of N-step feasible initial states. If δ in (6.33) or in (6.41) is chosen such that the Problems 1 or 2 are feasible, then X_f is a non-empty set. In parametric programming problems one seeks the solution $\mathbf{U}^*(\tilde{\mathbf{x}})$ as an explicit function of the parameters $\tilde{\mathbf{x}}$ in some set $X \subseteq X_f \subseteq \mathbb{R}^{\tilde{n}}$ [15].

Approximate mp-NLP Approach to Explicit GP-NMPC

In [19, 24], an approximate mp-NLP approach is proposed to explicitly solve the GP-NMPC problems formulated in the Sect. 6.3.2.3. Let $X \subset \mathbb{R}^{\tilde{x}}$ be a hyper-rectangle where we seek to approximate the optimal solution $\mathbf{U}^*(\tilde{\mathbf{x}})$ to Problem 3. It is required that the parameter space partition is orthogonal and can be represented as a $k - d$ tree. The idea of the approximate mp-NLP approach is to construct a piecewise linear (PWL) approximation $\hat{\mathbf{U}}(\tilde{\mathbf{x}})$ to $\mathbf{U}^*(\tilde{\mathbf{x}})$ on X, where the constituent linear functions are defined on hyper-rectangles covering X. The computation of a linear feedback $\hat{\mathbf{U}}_0(\tilde{\mathbf{x}})$, associated to a given region X_0, includes the following steps [19, 24]. First, a close-to-global solution of Problem 3 is computed at a set of points $V_0 = \{\mathbf{v}_0, \mathbf{v}_1, \ldots, \mathbf{v}_{N_1}\} \subset X_0$. Then, based on the solutions at these points, a local linear approximation $\hat{\mathbf{U}}_0(\tilde{\mathbf{x}}) = \mathbf{K}_0\tilde{\mathbf{x}} + \mathbf{g}_0$ to the close-to-global solution $\mathbf{U}^*(\tilde{\mathbf{x}})$, valid in the whole hyper-rectangle X_0, is determined by applying the procedure [19, 24]:

Procedure 1 (Computation of explicit approximate solution). *Consider any hyper-rectangle $X_0 \subseteq X$ with a set of points $V_0 = \{\mathbf{v}_0, \mathbf{v}_1, \mathbf{v}_2, \ldots, \mathbf{v}_{N_1}\} \subset X_0$. Compute $\mathbf{K}_0$ and $\mathbf{g}_0$ by solving the following NLP:*

$$\min_{\mathbf{K}_0, \mathbf{g}_0} \sum_{i=0}^{N_1} (J(\mathbf{K}_0\mathbf{v}_i + \mathbf{g}_0, \mathbf{v}_i) - V^*(\mathbf{v}_i) + \alpha\|\mathbf{K}_0\mathbf{v}_i + \mathbf{g}_0 - \mathbf{U}^*(\mathbf{v}_i)\|^2) \qquad (6.49)$$

$$\text{subject to } G(\mathbf{K}_0\mathbf{v}_i + \mathbf{g}_0, \mathbf{v}_i) \leq 0, \forall \mathbf{v}_i \in V_0 \qquad (6.50)$$

In (6.49), $J(\mathbf{K}_0\mathbf{v}_i + \mathbf{g}_0, \mathbf{v}_i)$ is the sub-optimal cost, $V^*(\mathbf{v}_i)$ denotes the cost corresponding to the close-to-global solution $\mathbf{U}^*(\mathbf{v}_i)$, i.e. $V^*(\mathbf{v}_i) = J(\mathbf{U}^*(\mathbf{v}_i), \mathbf{v}_i)$, and the parameter α is a weighting coefficient (tuned in an ad-hoc fashion). Note that the computed linear feedback $\hat{\mathbf{U}}_0(\tilde{\mathbf{x}}) = \mathbf{K}_0\tilde{\mathbf{x}} + \mathbf{g}_0$ satisfies the constraints in Problem 3 only at the discrete set of points $V_0 \subset X_0$. After the feedback $\hat{\mathbf{U}}_0(\tilde{\mathbf{x}})$ has been determined, an estimate $\hat{\varepsilon}_0$ of the maximal cost function approximation error in X_0 is computed as follows:

$$\widehat{\varepsilon}_0 = \max_{i \in \{0,1,2,\ldots,N_1\}} \left(J(\mathbf{K}_0 \mathbf{v}_i + \mathbf{g}_0, \mathbf{v}_i) - V^*(\mathbf{v}_i) \right) \tag{6.51}$$

If $\widehat{\varepsilon}_0 > \bar{\varepsilon}$, where $\bar{\varepsilon} > 0$ is the specified tolerance of the approximation error, the region X_0 is divided and the procedure is repeated for the new regions. The approximate PWL feedback law can be found by applying the approximate mp-NLP algorithms, described in [19]. These algorithms terminate with a PWL function $\widehat{\mathbf{U}}(\widetilde{\mathbf{x}}) = [\widehat{\mathbf{u}}(0, \widetilde{\mathbf{x}}), \widehat{\mathbf{u}}(1, \widetilde{\mathbf{x}}), \ldots, \widehat{\mathbf{u}}(N-1, \widetilde{\mathbf{x}})]$ that is defined on an inner approximation X_Π of the set $X \cap X_f$.

6.3.3 Adaptive Control

Adaptive controller is the controller that continuously adapts to some changing process. Adaptive controllers emerged in early sixties of the previous century. At the beginning these controllers were mainly adapting themselves based on linear models with changing parameters. Since then several authors have proposed the use of non-linear models as a base to build nonlinear adaptive controllers. These are meant for the control of time-varying nonlinear systems or of time-invariant nonlinear systems that are modelled as parameter-varying simplified nonlinear models.

Various divisions of adaptive control structures are possible. One possible division [30] is into open-loop and closed-loop adaptive systems.

Open-loop adaptive systems are gain-scheduling or parameter-scheduling controllers. Closed-loop adaptive systems can be further divided to dual and non-dual adaptive systems.

Dual adaptive systems [16, 63] are those where the optimisation of the information collection and the control action are pursued at the same time. The control signal should ensure that the system output cautiously tracks the desired reference value and at the same time excites the plant sufficiently to accelerate the identification process. The solution to the dual control problem is based on dynamic programming and the resulting functional equation is often the Bellman equation. Not a large number of such controllers have been developed.

The difficulties to find the optimal solution for dual adaptive control lead to suboptimal adaptive dual controllers [16, 63] obtained by either various approximations or by reformulating the problem. Such a reformulated adaptive dual control problem is when a special cost function is considered, which consists of two added parts: control losses and an uncertainty measure. This is appealing for application with the Gaussian process model that provides measures of uncertainty.

Many adaptive controllers in general are based on the separation principle [63] that implies separate estimation of system model, i.e., system parameters, and the application of this model for control design. When the identified model used for control design and adaptation is presumed to be the same as the true system then the adaptive controller of this kind is said to be based on certainty equivalence principle and such adaptive control is named non-dual adaptive control. The control actions

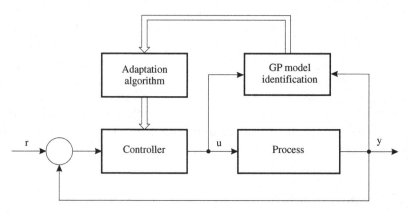

Fig. 6.7 General block scheme of the closed-loop system with adaptive controller

in non-dual adaptive control do not take any active actions that will influence the uncertainty.

When using the GP model for the adaptive control, different from gain-scheduling control, the GP model is identified on-line and this model is used in the control algorithm. The block scheme showing the general principle of adaptive control with the GP model identification is given in Fig. 6.7. It is sensible that advantages of GP models are considered in the control design, which relates the GP model-based adaptive control at least to suboptimal dual adaptive control principles. The uncertainty of model predictions obtained with the GP model prediction are dependent, among others, on local learning-data density, and the model complexity is automatically related to the amount and the distribution of the available data—more complex models need more evidence to make them likely. Both aspects are very useful in sparsely-populated transient regimes. Moreover, since weaker prior assumptions are typically applied in a nonparametric model, the bias is typically lower than in parametric models.

The above ideas are indeed related to the work done on the dual adaptive control, where the main effort has been concentrated on the analysis and design of adaptive controllers based on the use of the uncertainty associated with parameters of models with a fixed structure [16, 61].

The major differences in up-to-now published adaptive systems based on GP models are in the way how the on-line model identification is pursued.

Increasing the size of the covariance matrix, i.e., 'blow-up model', with the in-streaming data and repeating model optimisation is used in papers [47, 48, 60, 62] and [61], where more attention is devoted to control algorithms and their benefits based on information gained from the GP model and not on the model identification itself.

Another adaptive control algorithm implementation is control with feedback for cancelling nonlinearities, described already in Sect. 6.3.1 with the on-line learning of the inverse model. This sort of adaptive control with the increasing covariance matrix with the in-streaming data is described in [49]. Two sorts of on-line learning for the mentioned feedforward contained control is described in [52]. The first sort is with moving window strategy, where the old data are dropped from the on-line learned model, while the new data is accommodated, the second one accommodates only new data with sufficient information gain. These applications of referenced inverse GP models do not use entire information from the prediction distribution, but like those non-adaptive based on the same principle from Sect. 6.3.1 they focus on the mean value of prediction.

Alternatively to listed adaptive controllers, the adaptive control system principle described by [55] is based on the evolving GP model. The basic idea of the control based on the evolving system model is that the system GP model evolves with the in-streaming data and the information about system from the model is then used for its control. One option is that the information can be in the form of the GP model prediction for one or several steps ahead which is then used to calculate the optimal control input in the controlled system. Different possibilities exist for the evolving GP model depending on the level of changes we accommodate in the evolving system model. On the other hand, various control algorithms can be used depending on the GP model or closed-loop requirements.

A lot of GP model-based adaptive-control algorithms from the referenced publications are based on the Minimum Variance controller. One of the reasons is that the Minimum Variance controller explores the variance that is readily available with the GP model prediction.

The Minimum Variance controller in general [30] looks for a control signal $u(k)$ in time instant k, that will minimize the following cost function:

$$J_{MV} = E\{\| \mathbf{r}(k) - \mathbf{y}(k+m) \|^2\} \tag{6.52}$$

In this case, J_{MV} refers to the covariance of the error between the vector of set-points $\mathbf{r}(k)$ and the controlled outputs m-time steps in the future, $\mathbf{y}(k+m)$. The desired controller is thus the one that minimizes these variances, hence the name Minimum Variance control. The optimal control signal $\mathbf{u}_{opt}$ can be obtained by minimising selected cost function. The minimisation can be done analytically, but also numerically, using any appropriate optimisation method.

The cost function (6.52) can be expanded with a penalty terms and generalised to multiple-input multiple-output case leading to Generalized Minimum Variance control [62].

$$J_{GMV} = E\{\|\mathbf{r}(k+1) - \mathbf{y}(k+1)\|^2_{\mathbf{Q}}\} + \|\mathbf{u}_k\|^2_{\mathbf{R}} \tag{6.53}$$

where matrix $\mathbf{Q}$ is positive definite matrix and $\mathbf{R}$ is polynomial matrix with the backward shift operator q^{-1}. The matrix $\mathbf{Q}$ elements and matrix $\mathbf{R}$ polynomial coefficients can be used as tuning parameters.

The method named **Gaussian Process Dynamic Programming** (GPDP) is a Gaussian-process model-based adaptive control algorithm with the closest proximity to dual adaptive control. The details of the method are described in [13]. The following description is summarised from [13, 9]. The evolution of method can be followed with publications [8, 9, 13, 57, 58].

GPDP is an approximate dynamic programming method, where cost functions, so-called value functions in the dynamic programming recursion are modelled by GPs.

Reader is referred to [13] for details and demonstration of the method. Unfortunately, according to the method's authors [13], GPDP cannot be directly applied to a dynamic system, because it is often not possible to experience arbitrary state transitions. Moreover GPDP method does not scale that well to high dimensions.

More promising for engineering control applications is **Probabilistic Inference and Learning for Control** (PILCO) method, described in [10–12].

PILCO is a policy search method and an explicit value function model is not required as in GPDP method. The general idea of the method is to learn the system model with Reinforcement learning and control the closed-loop system, taking into account the probabilistic model of the process. The algorithm can be divided into three layers: a top level for the controller adaptation, an intermediate layer for the approximate inference for long-term predictions and a bottom layer for learning the model dynamics. A start state $\mathbf{x}_0$ is required by the algorithm in the beginning.

The PILCO method was applied to real systems, e.g., robotic systems [12].

6.4 Design of Explicit GP-NMPC of a Gas–Liquid Separator

6.4.1 The Gas–Liquid Separator

The semi-industrial process plant used for the case study separates gas from liquid in the context of technological waste-water treatment. The plant scheme and photo are shown in Figs. 6.8 and 6.9.

The role of the separation unit is to capture flue gases under low pressure from the effluent channels by means of water flow, to cool them down, and then supply them under high-enough pressure to other parts of the pilot plant. The flue gases coming from the effluent channels are absorbed by the water flow into the water circulation pipe through an injector.

The water flow is generated by the water ring pump. The speed of the pump is kept constant. The mixture of water and gas is pumped into the tank, where gas is separated from water. Hence, the accumulated gas in the tank forms a sort of 'gas cushion' with increased internal pressure. Owing to this pressure, the flue gas is blown out from the tank into the neutralization unit. On the other hand, the 'cushion' forces water to circulate back to the tank. The quantity of water in the circuit is constant.

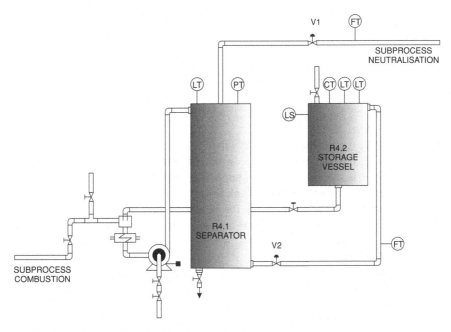

Fig. 6.8 The scheme of gas–liquid separator plant

Fig. 6.9 The photography of
gas–liquid separator plant

In order to understand the basic relations among variables and process nonlinearity, a first-principles model is employed [38]. The dynamics of air pressure in
the cushion (p) and the water level (h) are described by a set of two equations.

$$\frac{dp}{dt} = \frac{1}{S_1(h_{T_1} - h)} \left(p_0 (\alpha_0 + \alpha_1 p + \alpha_2 p^2 - k_1 R_1^{u_1 - 1} \sqrt{p}) \right.$$
$$\left. + (p_0 + p) \left(\Phi_w - k_2 R_2^{u_2 - 1} \sqrt{p + k_w(h - h_{T_2})} \right) \right) \tag{6.54}$$

$$\frac{dh}{dt} = \frac{1}{S_1} \left(\Phi_w - k_2 R_2^{u_2 - 1} \sqrt{p + k_w(h - h_{T_2})} \right) \tag{6.55}$$

where u_i is the command signal of valve Vi, $i = 1, 2$, where, following the notation in Fig. 6.8, V1 is the valve on output from tank R4.1 to another sub-process and V2 is the valve between tanks R4.2 and R4.1, h is the level in tank R4.1, p is the relative air pressure in tank R4.1, S_1 is the section area of tank R4.1, p_0 is the atmospheric pressure, h_{Ti} is the height of tank R4.i, $i = 1, 2$, R_i is the open-close flow ratio of valve Vi, $i = 1, 2$, k_i is the flow coefficient of valve Vi, $i = 1, 2$, Φ_w is the known constant water flow through the pump, and α_i, $i = 1, 2, 3$ are constant parameters.

The dynamics of the servo-valves V1 and V2 are that of typical three-position valves, i.e.

$$u_{i,LIM} = \begin{cases} v_{i,MAX} & if \quad u_i > v_{i,MAX} \\ 0 & if \quad u_i < 0 \\ u_i & otherwise \end{cases} \quad i = 1, 2 \tag{6.56}$$

$$\dot{v}_i = \begin{cases} \dot{v}_{i,MAX} & if \quad u_{i,LIM} > v_i \\ \dot{v}_{i,MIN} & if \quad u_{i,LIM} < v_i \\ 0 & if \quad u_{i,LIM} = v_i \end{cases} \quad i = 1, 2 \tag{6.57}$$

where $u_{i,LIM}$ is an intermediate variable.

$$v_i = u_i \quad i = 1, 2 \tag{6.58}$$

when the valve is not saturated. It can be seen from this valve model that a rate constraint is present at the system inputs. The equations in (6.54)–(6.55) show that this is a nonlinear model, which will result in different dynamic behaviour depending on the operating region. From the model presented, it can be seen that the nonlinear process is multivariable (two inputs and two outputs with dynamic interactions between the channels).

User-friendly experimentation with the process plant is enabled within the Matlab/Simulink environment with an interface that enables PLC access with the Matlab/Simulink using OPC protocol via TCP/IPv4 over Ethernet IEEE802.3. Control algorithms for experimentation are run in Matlab code or as Simulink blocks and extended with functions/blocks, which access the PLC. In our case, all schemes for data acquisition and control are realised as Simulink blocks.

6.4.2 Gaussian Process Model of the Gas–Liquid Separator

In [36], a Gaussian process model of the separator's dynamics has been obtained based on measurement data for the input and the output signals, sampled with sampling time of 20 s. The model is composed of two parts: one is the sub-model that predicts the pressure p and the other is the sub-model that predicts the liquid level h. We limited the selection of the possible functional parts in the covariance function to the following functions:

- Square exponential or Gaussian covariance function:

$$C_f(\mathbf{z}_r, \mathbf{z}_q) = v_1 \exp\left[-\frac{1}{2}\sum_{d=1}^{D} w_d(z_{dr} - z_{dq})^2\right], \qquad (6.59)$$

- Linear covariance function:

$$C_f(\mathbf{z}_r, \mathbf{z}_q) = \sum_{d=1}^{D} w_d z_{dr} z_{dq}, \qquad (6.60)$$

- Rational quadratic covariance function:

$$C_f(\mathbf{z}_r, \mathbf{z}_q) = v_1\left[1 + \frac{1}{2\alpha}\sum_{d=1}^{D} w_d(z_{dr} - z_{dq})^2\right]^{-\alpha}, \qquad (6.61)$$

The noise part $C_n(\mathbf{z}_r, \mathbf{z}_q)$ in the covariance function is supposed to correspond to a white noise and is given by:

$$C_n(\mathbf{z}_r, \mathbf{z}_q) = v_0 \delta_{rq} \qquad (6.62)$$

In (6.59)–(6.62), the meaning of the variables and parameters is as described in Sect. 6.2.

A systematical iterative procedure of comparing modelling results with performance measures for various combinations of covariance functions for functional part and of different input regressors for two output models was pursued [36]. The following regressor GP sub-models were obtained for the pressure:

$$\mathbf{M}_{p1}: \quad p(k+1) = f_{p1}(p(k), u_1(k), h(k)) \qquad (6.63)$$

$$\mathbf{M}_{p2}: \quad p(k+1) = f_{p2}(p(k), p(k-1), u_1(k), u_2(k), h(k)) \qquad (6.64)$$

and for the liquid level:

$$\mathbf{M}_{h1}: \quad h(k+1) = f_{h1}(h(k), u_2(k), p(k)), \qquad (6.65)$$

Table 6.1 Comparison of models fit for pressure output (*the left part of table*) and liquid level output (*the right part of table*)

Pressure output				Liquid level output			
Regressors (Cov. function)	Ident. $\mathscr{L}$	Valid. MRSE	Valid. LPD	Regressors (Cov. function)	Ident. $\mathscr{L}$	Valid. MRSE	Valid. LPD
$M_{p1}^{\clubsuit}$ (G)	−2,321	0.047	−0.947	$M_{h1}^{\clubsuit}$ (G)	−1,955	0.051	1.799
$M_{p1}^{\clubsuit}$ (RQ)	−2,328	0.046	−1.099	M_{h1} (RQ)	−1,955	0.051	1.758
M_{p1} (LIN)	−1,757	0.193	6.414	M_{h1} (LIN)	−1,949	0.514	491
M_{p2} (G)	−2,439	0.055	−0.041	M_{h2} (G)	−2,037	0.004	−3.293
M_{p2} (RQ)	−2,449	0.054	−0.093	$M_{h2}^{\clubsuit}$ (RQ)	−2,037	0.004	−3.293
M_{p2} (LIN)	−2,007	∞	2117	M_{h2} (LIN)	−2,013	0.010	−3.050

The best sub-models, which can be used for accurate predictions of the plant behaviour are marked with $\clubsuit$, while the sub-models used for the design of explicit stochastic NMPC are marked with $\clubsuit$

$$M_{h2}: \quad h(k+1) = f_{h2}(h(k), h(k-1), u_2(k), p(k)) \qquad (6.66)$$

where k denotes the time instance. The backward approach from the higher number of regressors towards the lower number was used. 727 input-output data pairs were sampled uniformly with sampling time of 20 s from input and identification output signal and used as estimation data for Gaussian process models. The same number of samples from input and output signals for validation was used as validation data for Gaussian process models' predictions.

Validation measures for a sample of cases at the end of systematic selection of model structure is given in Table 6.1. The table gives logarithm of likelihood $\mathscr{L}$ that is used as a validation measure on identification data and MRSE and LPD measures for validation of predictions on validation input signal. The upper part of the table consists of the first-order models M_{p1} and M_{h1} with Gaussian (G), rational quadratic (RQ) and linear (LIN) covariance function. The lower part includes the second-order models M_{p2} and M_{h2} with the same three forms of covariance functions. In selecting the best sub-models, which can be used for accurate predictions of the plant behaviour, higher weight was put on the LPD measure, because it is more appropriate for Bayesian models as it incorporates also the variance of predictions. Thus, for the purpose of accurate predictions of output variables based on input signals, the winning models are $M_{p1}^{\clubsuit}$ and $M_{h2}^{\clubsuit}$, which are based on rational quadratic covariance function (6.61) for the functional part of the covariance function (6.62). These two models give better validation results with validation signals than models with other covariance functions and other regressors. However, here the objective is to design an explicit stochastic NMPC controller for the gas–liquid separator. It is known that the partition complexity and the memory storage requirements, associated with the explicit MPC controllers, increase rapidly with the increase of the parameters space dimension [18]. Thus, if the models $M_{p1}^{\clubsuit}$ and $M_{h2}^{\clubsuit}$ are used, then the parameters space would be 3-dimensional (defined by the

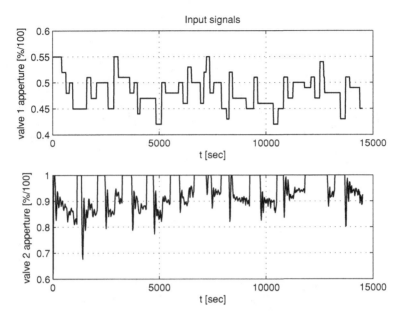

Fig. 6.10 Validating input signals

regressors $p(k), h(k), h(k-1))$, while the parameters space would be 2-dimensional if the models $\mathrm{M}_{p1}^{\blacklozenge}$ and $\mathrm{M}_{h1}^{\blacklozenge}$ are used (in this case the regressors are $p(k)$ and $h(k)$). In order to obtain an explicit stochastic NMPC with less complexity, we have chosen to use the models $\mathrm{M}_{p1}^{\blacklozenge}$ and $\mathrm{M}_{h1}^{\blacklozenge}$. These sub-models have a Gaussian covariance function, expressed as:

$$C^j(\mathbf{z}^j(r), \mathbf{z}^j(q)) = v_1^j \exp\left[-\frac{1}{2}\sum_{d=1}^{D} w_d^j (z_d^j(r) - z_d^j(q))^2\right] + v_0^j \delta_{rq}, \ j = 1, 2, \quad (6.67)$$

where $j = 1$ is associated to the model (6.63), $j = 2$ corresponds to the model (6.65), $D = 3$ is the number of input signals in both models, r and q are discrete time instances, and δ_{rq} is the Kronecker operator. Thus, the model (6.63) has covariance function $C^1(\mathbf{z}^1, \mathbf{z}^1)$ (where $\mathbf{z}^1 = [p, u_1, h]$) with the following parameters:

$$[w_1^1, w_2^1, w_3^1, v_0^1, v_1^1] = [1.71, 0.10, 1.68, 8.26 \times 10^{-5}, 0.27] \quad (6.68)$$

and the model (6.65) has covariance function $C^2(\mathbf{z}^2, \mathbf{z}^2)$ (where $\mathbf{z}^2 = [h, u_2, p]$) with the following parameters:

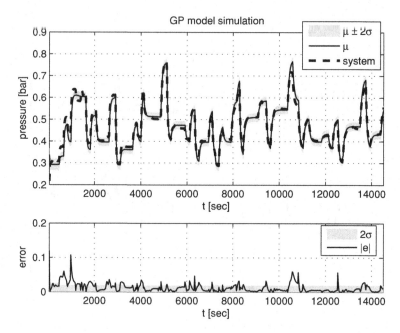

Fig. 6.11 Comparison of the pressure measurement and the Gaussian process model $(M_{p1}^\blacklozenge)$ response

$$[w_1^2, w_2^2, w_3^2, v_0^2, v_1^2] = [4.81, 97.19, 848.51, 2.55 \times 10^{-4}, 7.28] \qquad (6.69)$$

The simulation responses on the validation input signals (Fig. 6.10), that was different from the identification one, of the models $M_{p1}^\blacklozenge$ and $M_{h1}^\blacklozenge$ are given in Figs. 6.11 and 6.12. The simulation responses of the models $M_{p1}^\clubsuit$ and $M_{h2}^\clubsuit$ on the same validation signals are given in Figs. 6.13 and 6.14.

6.4.3 Design and Performance of Explicit GP-NMPC

Based on the Gaussian process model (6.63) and (6.65), an explicit stochastic NMPC controller for the gas–liquid separation plant is designed. For this purpose, the approximate mp-NLP approach in [24] is applied. The following constraints are imposed on the control inputs:

$$0 \le u_1(k) \le 1, \; 0 \le u_2(k) \le 1 \qquad (6.70)$$

The set point values for the pressure and the liquid level are:

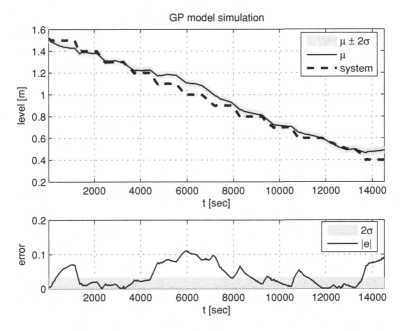

Fig. 6.12 Comparison of the liquid level measurement and the Gaussian process model ($M_{h1}^{\blacklozenge}$) response

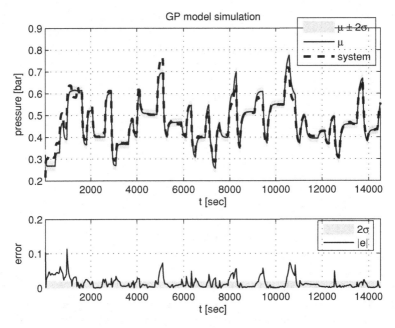

Fig. 6.13 Comparison of the pressure measurement and the Gaussian process model ($M_{p1}^{\clubsuit}$) response

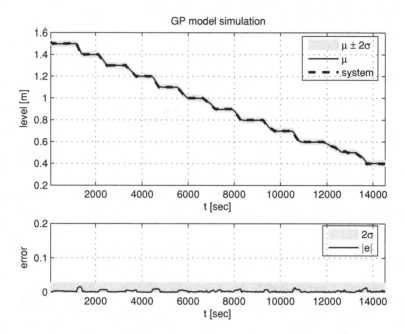

Fig. 6.14 Comparison of the liquid level measurement and the Gaussian process model ($M_{h2}^{\clubsuit}$) response

$$r_p = 0.5\,[\text{bar}], \quad r_h = 1.4\,[\text{m}] \tag{6.71}$$

The model steady state values of the two control inputs corresponding to these set point values are:

$$u_{1_{r_p}} = 0.47, \quad u_{2_{r_h}} = 0.848 \tag{6.72}$$

The aim is to minimize the cost function:

$$J(\mathrm{U}, \mathbf{x}(k)) = \sum_{j=0}^{N-1} \left[\| \, \mu(\mathbf{x}(k+j|k)) - \mathbf{r} \|_{\mathbf{Q}}^2 + \| \mathbf{u}(k+j) - \mathbf{u}_r \|_{\mathbf{R}}^2 \right] \\ + \| \, \mu(\mathbf{x}(k+N|k)) - \mathbf{r} \, \|_{\mathbf{P}}^2 \tag{6.73}$$

with $\mathrm{U} = [\mathbf{u}(k|k), \mathbf{u}(k+1|k), \ldots, \mathbf{u}(k+N-1|k)]$, $\mathbf{x} = [p, h]$, $\mathbf{r} = [r_p, r_h]$, $\mathbf{u} = [u_1, u_2]$, $\mathbf{u_r} = [u_{1_{r_p}}, u_{2_{r_h}}]$, subject to the inputs constraints (6.70) and the Gaussian process model (6.63) and (6.65). The horizon is $N = 5$ and the weighting matrices are $\mathbf{Q} = \mathbf{P} = \mathrm{diag}\{1, 200\}$, $\mathbf{R} = \mathrm{diag}\{0.5, 0.5\}$.

In [31], a condition on the tolerance of the cost function approximation error has been derived such that the asymptotic stability of the nonlinear system in closed-loop with the approximate explicit NMPC is guaranteed. According to this

Fig. 6.15 State space
partition of the explicit
approximate stochastic
NMPC controller and the
approximate (*solid curve*) and
the exact (*dotted curve*) state
trajectories

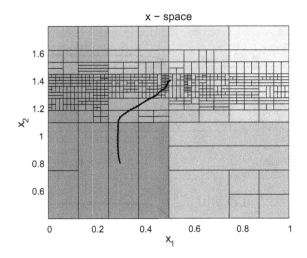

Fig. 6.16 The suboptimal
control functions

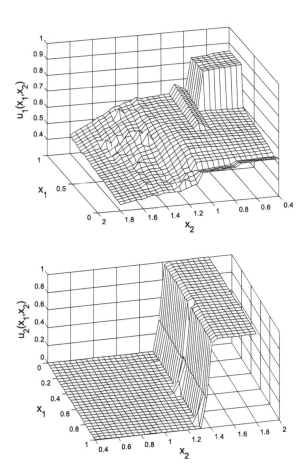

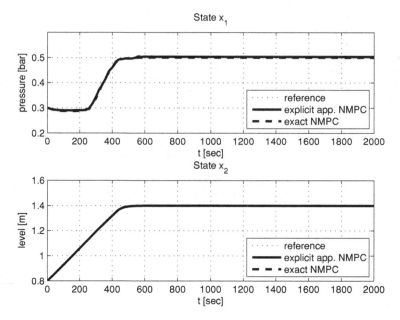

Fig. 6.17 The pressure x_1 and the liquid level x_2

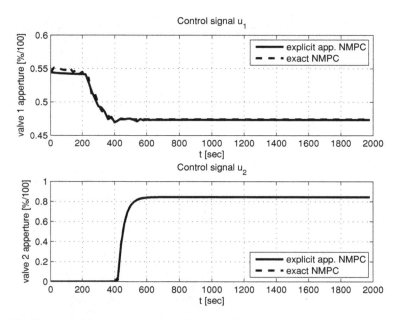

Fig. 6.18 The control inputs (the aperture of valves V1 and V2)

condition, the tolerance is chosen to be dependent on the state, which would lead to a state-space partition with less complexity in comparison to that corresponding to an uniform tolerance. Here, a similar approach is applied and the tolerance is chosen to be $\bar{\varepsilon}(X_0) = \max(\bar{\varepsilon}_a, \bar{\varepsilon}_r \min_{\mathbf{x} \in X_0} V^*(\mathbf{x}))$, where $\bar{\varepsilon}_a = 0.005$ and $\bar{\varepsilon}_r = 0.1$ are the absolute and the relative tolerances. Here, $X_0 \subset X$, where $X = [0, 1] \times [0.4, 1.8]$ is the state space to be partitioned. The state-space partition of the explicit stochastic NMPC controller is shown in Fig. 6.15. The partition has 487 regions and 13 levels of search. Totally, 21 arithmetic operations are needed in real-time to compute the control input (13 comparisons, 4 multiplications, and 4 additions). In Fig. 6.16, the suboptimal control functions, associated with the explicit approximate stochastic NMPC controller, are shown. The performance of the closed-loop system was simulated for initial state $\mathbf{x}(0) = [0.3\ 0.8]^T$. The response is depicted in the state space (Fig. 6.15), as well as trajectories in time (Figs. 6.17, 6.18).

6.5 Conclusions

In this chapter a review of some of the methods for modelling and control of complex stochastic systems based on Gaussian process models is given. The Gaussian process regression modelling method is a computational intelligence method that can be applied for dynamic systems identification. The resulting models can be further used for various sorts of control design.

In this chapter an overview of inverse dynamics control, model-based predictive control and adaptive control based on Gaussian process models is given. An approximate mp-NLP approach to explicit solution of reference tracking NMPC problems based on Gaussian process models is developed to demonstrate a control of a semi-industrial gas–liquid separation plant.

Simulations of the closed-loop system show the high quality performance of the approximate explicit stochastic NMPC controller. Although the obtained results are based on simulation data, the case-study shows the potential use of the considered approach in the industrial practice.

References

1. Alessio, A., Bemporad, A.: A survey on explicit model predictive control. In: Magni, L., Raimondo, D.M., Allgöwer, F. (eds.) Nonlinear Model Predictive Control: Towards New Challenging Applications. Lecture Notes in Control and Information Sciences, vol. 384, pp. 345–369. Springer, Berlin (2009)
2. Ažman, K., Kocijan, J.: Application of Gaussian processes for black-box modelling of biosystems. ISA Trans. **46**, 443–457 (2007)
3. Ažman, K., Kocijan, J.: Non-linear model predictive control for models with local information and uncertainties. Trans. Inst. Meas. Control **30**(5), 371–396 (2008)

4. Bemporad, A., Morari, M., Dua, V., Pistikopoulos, E.N.: The explicit linear quadratic regulator for constrained systems. Automatica **38**, 3–20 (2002)
5. Cannon, M., Couchman, P., Kouvaritakis, B.: MPC for stochastic systems. In: Findeisen, R., Allgöwer, F., Biegler, L.T. (eds.) Assessment and Future Directions of Nonlinear Model Predictive Control. Lecture Notes in Control and Information Sciences, vol. 358, pp. 255–268. Springer, Berlin (2007)
6. Couchman, P., Cannon, M., Kouvaritakis, B.: Stochastic MPC with inequality stability constraints. Automatica **42**, 2169–2174 (2006)
7. Couchman, P., Kouvaritakis, B., Cannon, M.: LTV models in MPC for sustainable development. Int. J. Control **79**, 63–73 (2006)
8. Deisenroth, M., Peters, J., Rasmussen, C.: Approximate dynamic programming with Gaussian processes. In: Proceedings of American Control Conference (ACC), pp. 4480–4485. Seattle, WA (2008)
9. Deisenroth, M., Rasmussen, C.: Bayesian inference for efficient learning in control. In: Proceedings of Multidisciplinary Symposium on Reinforcement Learning (MSRL). Montreal, Canada (2009)
10. Deisenroth, M., Rasmussen, C.: Efficient reinforcement learning for motor control. In: Proceedings of the 10th International Ph.D. Workshop on Systems and Control: A Young Generation Viewpoint. Hluboka nad Vltavou, Czech Republic (2009)
11. Deisenroth, M., Rasmussen, C., Peters, J.: Model-based reinforcement learning with continuous states and actions. In: Proceedings of the European Symposium on Artificial Neural Networks (ESANN), pp. 19–24. Bruges, Belgium (2008)
12. Deisenroth, M.P.: Efficient reinforcement learning using gaussian processes. Ph.D. thesis, Karlsruhe Institute of Technology, Karlsruhe (2010)
13. Deisenroth, M.P., Rasmussen, C.E., Peters, J.: Gaussian process dynamic programming. Neurocomputing **72**(7–9), 1508–1524 (2009)
14. Engel, Y., Szabo, P., Volkinshtein, D.: Learning to control an Octopus arm with Gaussian process temporal difference methods. In: Weiss, Y., Schoelkopf, B., Platt, J. (eds.) Advances in Neural Information Processing Systems, vol. 18, pp. 347–354. MIT Press, Cambridge (2006)
15. Fiacco, A.V.: Introduction to sensitivity and stability analysis in nonlinear programming. Academic Press, New York (1983)
16. Filatov, N., Unbehauen, H.: Survey of adaptive dual control methods. IEE Proc. Control Theory Appl. **147**(1), 119–128 (2000)
17. Girard, A., Murray-Smith, R.: Gaussian processes: prediction at a noisy input and application to iterative multiple-step ahead forecasting of time-series. In: Murray-Smith, R., Shorten, R. (eds.) Switching and Learning in Feedback Systems. Lecture Notes in Computer Science, vol. 3355, pp. 158–184. Springer, Berlin (2005)
18. Grancharova, A., Johansen, T.A.: Approaches to explicit nonlinear model predictive control with reduced partition complexity. In: Proceedings of European Control Conference, pp. 2414–2419. Budapest, Hungary (2009)
19. Grancharova, A., Johansen, T.A.: Explicit Nonlinear Model Predictive Control: Theory and Applications. Lecture Notes in Control and Information Sciences, vol. 429. Springer, Berlin (2012)
20. Grancharova, A., Johansen, T.A., Tøndel, P.: Computational aspects of approximate explicit nonlinear model predictive control. In: Findeisen, R., Allgöwer, F., Biegler, L.T. (eds.) Assessment and Future Directions of Nonlinear Model Predictive Control. Lecture Notes in Control and Information Sciences, vol. 358, pp. 181–190. Springer, Berlin (2007)
21. Grancharova, A., Kocijan, J.: Stochastic predictive control of a thermoelectric power plant. In: Proceedings of the International Conference Automatics and Informatics '07, pp. I-13–I-16. Sofia (2007)
22. Grancharova, A., Kocijan, J.: Explicit stochastic model predictive control of gas–liquid separator based on Gaussian process model. In: Proceedings of the International Conference on Automatics and Informatics, pp. B-85–B-88. Sofia, Bulgaria (2011)

23. Grancharova, A., Kocijan, J., Johansen, T.A.: Explicit stochastic nonlinear predictive control based on Gaussian process models. In: Proceedings of European Control Conference (ECC), pp. 2340–2347. Kos, Greece (2007)
24. Grancharova, A., Kocijan, J., Johansen, T.A.: Explicit stochastic predictive control of combustion plants based on Gaussian process models. Automatica **44**(4), 1621–1631 (2008)
25. Gregorčič, G., Lightbody, G.: Gaussian processes for internal model control. In: Rakar, A. (ed.) Proceedings of 3rd International Ph.D. Workshop on Advances in Supervision and Control Systems, a Young Generation Viewpoint, pp. 39–46. Strunjan, Slovenia (2002)
26. Gregorčič, G., Lightbody, G.: From multiple model networks to the Gaussian processes prior model. In: Proceedings of IFAC ICONS Conference, pp. 149–154. Faro (2003)
27. Gregorčič, G., Lightbody, G.: Internal model control based on Gaussian process prior model. In: Proceedings of the 2003 American Control Conference, ACC 2003, pp. 4981–4986. Denver, CO (2003)
28. Gregorčič, G., Lightbody, G.: Gaussian process approaches to nonlinear modelling for control. In: Intelligent Control Systems Using Computational Intelligence Techniques, IEE Control Series, vol. 70, pp. 177–217. IEE, London (2005)
29. Gregorčič, G., Lightbody, G.: Gaussian process internal model control. Int. J. Syst. Sci. 1–16 (2011). http://www.tandfonline.com/doi/abs/10.1080/00207721.2011.564326
30. Isermann, R., Lachman, K.H., Matko, D.: Adaptive Control Systems. Systems and Control Engineering. Prentice Hall International, New York (1992)
31. Johansen, T.A.: Approximate explicit receding horizon control of constrained nonlinear systems. Automatica **40**, 293–300 (2004)
32. Ko, J., Klein, D.J., Fox, D., Haehnel, D.: Gaussian processes and reinforcement learning for identification and control of an autonomous blimp. In: Proceedings of the International Conference on Robotics and Automation, pp. 742–747. Rome (2007)
33. Kocijan, J.: Control algorithms based on Gaussian process models: a state-of-the-art survey. In: Proceedings of the Special International Conference on Complex Systems: Synergy of Control, Communications and Computing—COSY 2011 (2011)
34. Kocijan, J.: Dynamic GP models: an overview and recent developments. In: Recent Researches in Applied Mathematics and Economics: Proceedings of the 6th International Conference on Applied Mathematics, Simulation, Modelling, (ASM'12), pp. 38–43. Vougliameni, Greece (2012)
35. Kocijan, J., Girard, A., Banko, B., Murray-Smith, R.: Dynamic systems identification with Gaussian processes. Math. Comput. Modell. Dyn. Syst. **11**(4), 411–424 (2005)
36. Kocijan, J., Grancharova, A.: Gaussian process modelling case study with multiple outputs. C R Acad Bulg. Sci. **63**(4), 601–608 (2010)
37. Kocijan, J., Leith, D.J.: Derivative observations used in predictive control. In: Proceedings of IEEE Melecon Conference, vol. 1, pp. 379–382. Dubrovnik (2004)
38. Kocijan, J., Likar, B.: Gas–liquid separator modelling and simulation with Gaussian-process models. Simul. Model. Pract. Theory **16**(8), 910–922 (2008)
39. Kocijan, J., Murray-Smith, R.: Nonlinear predictive control with Gaussian process model. In: Murray-Smith, R., Shorten, R. (eds.) Switching and Learning in Feedback Systems. Lecture Notes in Computer Science, vol. 3355, pp. 185–200. Springer, Heidelberg (2005)
40. Kocijan, J., Murray-Smith, R., Rasmussen, C., Girard, A.: Gaussian process model based predictive control. In: Proceedings of 4th American Control Conference (ACC 2004), pp. 2214–2218. Boston, MA (2004)
41. Kocijan, J., Murray-Smith, R., Rasmussen, C.E., Likar, B.: Predictive control with Gaussian process models. In: Proceedings of IEEE Region 8 EUROCON 2003: Computer as a Tool, vol. A, pp. 352–356. Ljubljana (2003)
42. Kouvaritakis, B., Cannon, M., Couchman, P.: MPC as a tool for sustainable development integrated policy assessment. IEEE Trans. Autom. Control **51**, 145–149 (2006)
43. Kraft, D.: On converting optimal control problems into nonlinear programming problems. In: Schittkowski, K. (ed.) Computational Mathematical Programming, NATO ASI Series, vol. F15, pp. 261–280. Springer, Berlin (1985)

44. Likar, B., Kocijan, J.: Predictive control of a gas–liquid separation plant based on a Gaussian process model. Comput. Chem. Eng. **31**(3), 142–152 (2007)
45. Maciejowski, J.M.: Predictive Control with Constraints. Pearson Education Limited, Harlow (2002)
46. Mayne, D.Q., Rawlings, J.B., Rao, C.V., Scokaert, P.O.M.: Constrained model predictive control: stability and optimality. Automatica **36**, 789–814 (2000)
47. Murray-Smith, R., Sbarbaro, D.: Nonlinear adaptive control using nonparametric Gaussian process prior models. In: Proceedings of IFAC 15th World Congress. Barcelona (2002)
48. Murray-Smith, R., Sbarbaro, D., Rasmussen, C., Girard, A.: Adaptive, cautious, predictive control with Gaussian process priors. In: Proceedings of 13th IFAC Symposium on System Identification. Rotterdam, Netherlands (2003)
49. Nguyen-Tuong, D., Peters, J.: Learning robot dynamics for computed torque control using local Gaussian processes regression. In: Symposium on Learning and Adaptive Behaviors for Robotic Systems, pp. 59–64 (2008)
50. Nguyen-Tuong, D., Peters, J., Seeger, M., Schoelkopf, B.: Learning inverse dynamics: a comparison. In: Proceedings of the European Symposium on Artificial Neural Networks (ESANN), pp. 13–18. Bruges, Belgium (2008)
51. Nguyen-Tuong, D., Seeger, M., Peters, J.: Computed torque control with nonparametric regression models. In: Proceedings of the 2008 American Control Conference, ACC 2008, p. 6. Seattle, Washington (2008)
52. Nguyen-Tuong, D., Seeger, M., Peters, J.: Real-time local GP model learning, vol. 264. From Motor Learning to Interaction Learning in Robots, pp. 193–207. Springer, Heidelberg (2010)
53. Norgaard, M., Ravn, O., Poulsen, N.K., Hansen, L.K.: Neural Networks for Modelling and Control of Dynamic Systems: A Practitioner's Handbook. Advanced Textbooks in Control and Signal Processing. Springer, London (2000)
54. Palm, R.: Multiple-step-ahead prediction in control systems with Gaussian process models and TS-fuzzy models. Eng. Appl. Artif. Intell. **20**(8), 1023–1035 (2007)
55. Petelin, D., Kocijan, J.: Control system with evolving Gaussian process model. In: Proceedings of IEEE Symposium Series on Computational Intelligence, SSCI 2011. IEEE, Paris (2011)
56. Pistikopoulos, E.N., Georgiadis, M.C., Dua, V.: Multi-parametric model-based control. Wiley-VCH, Weinheim (2007)
57. Rasmussen, C.E., Deisenroth, M.P.: Probabilistic inference for fast learning in control. In: Recent Advances in Reinforcement Learning. Lecture Notes on Computer Science, vol. 5323, pp. 229–242. Springer, Berlin (2008)
58. Rasmussen, C.E., Kuss, M.: Gaussian processes in reinforcement learning. In: S. Thurn, L. Saul, B. Schoelkopf (eds.) Advances in Neural Information Processing Systems Conference, vol. 16, pp. 751–759. MIT Press, Cambridge (2004)
59. Rasmussen, C.E., Williams, C.K.I.: Gaussian Processes for Machine Learning. MIT Press, Cambridge (2006)
60. Sbarbaro, D., Murray-Smith, R.: An adaptive nonparametric controller for a class of nonminimum phase non-linear system. In: Proceedings of IFAC 16th World Congress. Prague, Czech Republic (2005)
61. Sbarbaro, D., Murray-Smith, R.: Self-tuning control of nonlinear systems using Gaussian process prior models. In: Murray-Smith, R., Shorten, R. (eds.) Switching and Learning in Feedback Systems. Lecture Notes in Computer Science, vol. 3355, pp. 140–157. Springer, Heidelberg (2005)
62. Sbarbaro, D., Murray-Smith, R., Valdes, A.: Multivariable generalized minimum variance control based on artificial neural networks and Gaussian process models. In: International Symposium on Neural Networks. Springer, New York (2004)
63. Wittenmark, B.: Adaptive dual control. In: Control Systems, Robotics and Automation, Encyclopedia of Life Support Systems (EOLSS), Developed Under the Auspices of the UNESCO. Eolss Publishers, Oxford (2002)

Chapter 7
Computational Intelligence Techniques for Chemical Process Control

N. Paraschiv, M. Oprea, M. Cărbureanu and M. Olteanu

Abstract The chapter focuses on two computational intelligence techniques, genetic algorithms and neuro-fuzzy systems, for chemical process control. It has three sub-chapters: 1. Objectives and Conventional Automatic Control of Chemical Processes 2. Computational Intelligence Techniques for Process Control 3. Case study. A case study is described in detail that describes a neuro-fuzzy control system for a wastewater pH neutralization process.

7.1 Objectives and Conventional Automatic Control of Chemical Processes

The chemical industry represented and continues to represent a dynamical division of the world economy. On the terms of competitive markets, the justification of this dynamics is conferred by the fact that the products provided by the chemical industry represent basic materials to a number of other industries.

It is important to underline that producing conventional or renewable energy, computers and means of communications would not be possible aside from the existence of the products provided by the chemical industry. We can add amongst the essential products of the chemical industry fuels, medicines and different type of plastic.

N. Paraschiv · M. Oprea · M. Cărbureanu · M. Olteanu (✉)
Petroleum-Gas University, Ploieşti, Romania
e-mail: molteanu@upg-ploiesti.ro

N. Paraschiv
e-mail: nparaschiv@upg-ploiesti.ro

M. Oprea
e-mail: mihaela@upg-ploiesti.ro

M. Cărbureanu
e-mail: mcarbureanu@upg-ploiesti.ro

Although the ecologists are skeptic regarding the chemical industry, we have to underline the fact that this industry offers new perspectives in respect of processing waste products, obtaining non-pollutant fuels, producing biodegradable plastic, etc. In the first part of this section we will identify the objectives of chemical processes and the necessity of controlling these processes and in the next sections we will approach aspects concerning the conventional automatic control of some categories of chemical processes.

7.1.1 Objectives of Chemical Processes

Frequently met processes in the chemical industry are those associated to the phenomena of transfer and to chemical reactions. In the category of transfer processes, there are classified the ones of mass, thermal energy and impulse transfer. Known as unitary processes, these develope in specific installations, such as fractionating columns, heating furnaces, heat exchangers, condensers, reboilers, chemical reactors, gas compressors.

In chemical plants there are complex processes whose finality is represented by products used as such or which constitute basic materials for other plants. Irrespective of the character of a (complex or unitary) process, this does not represent a goal by itself but it is subordinated to some objectives, amongst which representatives are the ones of *quality*, *efficiency* and *security*.

The *quality* objectives are presented as specifications such as, for example, the compositions of separated products in the case of mass transfer, the temperatures of heated (cooled) products in the case of thermal transfer or the conversion degree of reactants associated to the chemical reactions.

Regarding the objectives of *efficiency*, these refer to the profitability of the process, respectively to the existence of a positive difference between the income obtained from the sale of a chemical product and the costs, necessary for its production. Concerning the objectives of *security*, these imply the deployment of the process so that the safety of the people, of the environment or of the related facilities should not be affected.

The control of a process implies the supervision of a process so that the objectives imposed to it should be achieved. If, for the design of a process, the objectives represent *starting points*, in the case of the control, these are targets.

The automatic control is based on the following functions of automation: *automatic monitoring, automatic control, automatic optimization, automatic safety*.

The function of *monitoring* offers the possibility of identifying the state of a process. Practically, monitoring implies the determination of the values of the parameters associated to a process, which can be achieved by measurement and/or by computation. In Fig. 7.1 it is represented a hierarchical approach, in which the inferior level (level 1) includes systems of measurement and the superior level (level 2), the relations for calculating the values of the parameters which are not

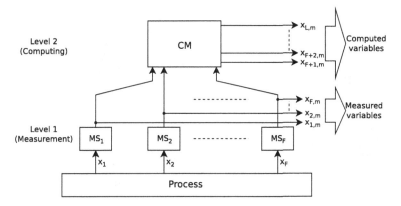

Fig. 7.1 Hierarchical monitoring system x_i process parameters, x_{im} measurement results, x_{ic} computed variables, MS_i Measurement System, CM Computing Module

measured. Normally, the number of parameters which are measured represents the number of the *degrees of freedom* of the respective process [1].

With reference to the function of *control*, a process is considered adjustable if it can be brought and maintained in a state of reference. Reaching and/or maintaining the state of reference imply the application of commands to the process.

From the point of view of complexity, the *automatic control systems* (ACS) can be conventional or advanced. A *conventional* ACS is usually associated to a single parameter. Based on the manner of action, *corrective* or *preventive*, these ACS can be: *feedback systems* (effect), respectively *feedforward systems* (cause), considered fundamental types of ACS, specific to the level of conventional automation.

The functioning of a conventional ACS dictates the existence of the functions of *measurement, command, execution*, achieved, in order, by *transducers, controllers* and *final control elements FCE*. Usually, the three elements are considered grouped in the *automatic device* (AD). Thus, it can be considered that, from a structural point of view, an ACS is composed of *AD* and *Process*. Another approach, concerning the structure, identifies at the level of an ACS a fixed part and a variable one. The fixed part includes the *process*, the *transducer* (for a feedforward ACS the *transducers*) and the *FCE*, while the variable part is represented by the *controller*.

Figure 7.2 presents a hierarchical structure of conventional control, in which at the inferior level it is present feedback control of the parameter y_1 and at the superior one the feedforward control of the parameter y_2.

Evolved structured ACS (*advanced control*) have associated extended objectives at the entire process. In the case of extended objectives, the controlled variables can be represented by synthetic parameters whose values are determined by computation. The advanced control does not exclude the conventional control, the two categories coexisting within the hierarchical control systems. The functions of monitoring are implied in the achievement of the quality objective and partially, of the security one.

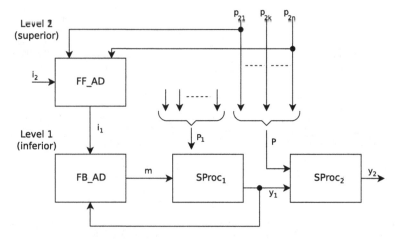

Fig. 7.2 Hierarchical conventional structure: *SProc1, SProc2* subprocesses, *FF_AD* Feedforward Automatic Device, *FB_AD* Feedback Automatic Device

After it has been indicated, among the objectives of a process there is the one which refers firstly to the protection of the human factor and of the environment, to the emergence of some events generated by an abnormal evolution of the process. The *Automatic Systems of Protection* (ASP), which can have *information functions* and/or *intervention functions*, assure the achievement of this objective. The functions of correct information are specific to the *Automatic Warning Systems* (AWS) and the ones of intervention are achieved by means of *Automatic Blocking Systems* (ABS) and *Automatic Systems of Command* (ASC).

AWS have an open structure and have also the role of informing the personnel implied in supervising and operating the process about the momentary state of a plant or about the apparition of an event. ABS assures the supervised removal of a plant or section of a plant from functioning, whereas it has not been intervened duly after the warning of prevention. Practically, the removal from functioning implies the blockage of supplying with energy and/or with raw material. A peculiarity of ABS is represented by the fact that these operate only when removed from functioning and not when reconnected. ACS are open systems, components of ASP, which assure the conditioned start of some plants or their normal stop (not in case of a breakdown). The conditioning of the start infers the authorization of reconnecting a facility only after it is observed the achievement of all the specified conditions.

All the three types of APS contain a *Logical Block of Command* (LBC) to whose level the logical functions which describe the sequences associated to the warning, blocking and command programs are implemented. LBC from the current APS have the related programs implemented exclusively in a programmed type of logic.

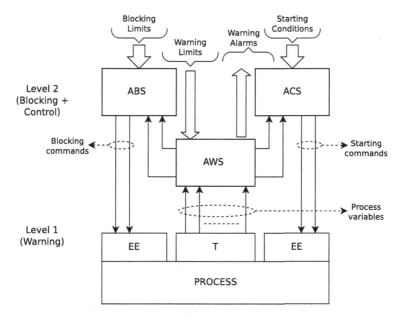

Fig. 7.3 Hierarchical structure of an AWS (Automatic Warning System): *ABS* Automatic Blocking System, *AWS* Automatic Warning System, *ACS* Automatic Command System

Beginning with the functions carried out in a SAP, we can assign to this a hierarchical structure on two levels, illustrated in Fig. 7.3, in which, at the first level, we find the informational systems, represented by AWS and at the second level there are the systems of intervention, represented by ABS and ACS.

The optimal control supposes the application to the process of those commands which bring to extreme an objective function (criterion or function of performance). The optimal commands are obtained by means of solving a problem of optimization which also assumes, besides the objective function, the existence of a method of searching the optimum, usually in the presence of some restrictions.

By optimal control, it is assured the achievement of the objective of efficiency, the objective function (functions) usually having an economical element or with economical implications.

The optimal control is situated at a hierarchical level, superior to the conventional automation, receiving values of information from this level and sending values of coordination to this one.

A particular case of control by fixing the values of reference is the one in which these values appear as a result of solving a problem of optimization. In the Fig. 7.4 there is represented the hierarchical structure organized on two levels, in which the first level concerns the conventional automation and the second one the optimization.

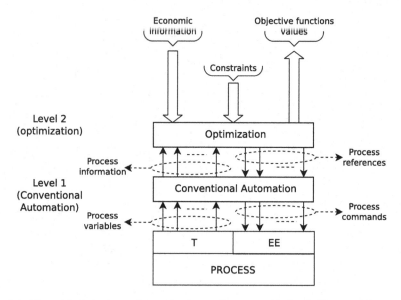

Fig. 7.4 Hierarchical structure of an optimal control system with fixed reference values

7.1.2 Conventional Automatic Control of Fractionating Processes

The fractionation represents one of the separation methods of a mixture in components or in groups of components. The main objective of a fractionation process is represented by conformation to the quality specifications for the products obtained by fractionation, which can be quantified in their compositions.

In the case of a fractionation column with a single feed flow rate, made up of n components and without side draw, the minimum number of parameters which must be measured for a complete knowledge of the column state is $F = (n + 2) + 10$ [1].

Regarding the control, for the exact conformation to the quality specifications, we should control the compositions on each tray of the fractionating column. Considering that the internal liquid and steam fluxes between the plates are not accessible, only the compositions of the products extract from the column can be controlled. Under these conditions, in case of a fractionating column without lateral fractions, there can be controlled only the compositions at the top of the column (distilled) respectively at its bottom (residue).

Another important parameter, determining for the fractionation is the pressure at which this process occurs. Thus, it is necessary also the control of this parameter. Beside the compositions and the column pressure, it is also necessary the control of the liquid stock on each tray, in the base of the column and in the reflux drum. Of reasons similar to the ones highlighted at the control of the compositions, it appears that only the accumulations of liquid (the stocks) from

Fig. 7.5 Controlled variables and manipulated variables for a fractionating column, x_D, x_B the concentration of light component in the *top* product and *bottom* product

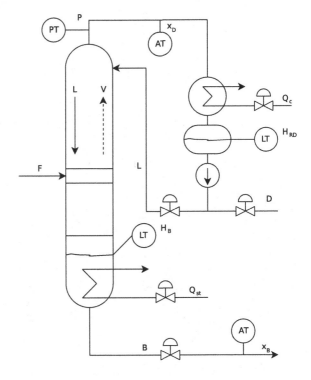

the base of the column and the from the reflux drum can be controlled. These are indirectly controlled, by means of the levels of the liquid H_B in the base of the column and H_{RD} in the reflux drum.

The control of a parameter is possible only for a fractionation column with a single feed flow rate and without intermediate products, the five control agents can be: the outputs of the products extracted from the column D (distilled) and B (residue), the reflux flow rate L and flow rates of the heating agent Q_{st} and cooling agent Q_c. In the Fig. 7.5 the five parameters which must be controlled have associated transducers (T) and to the five available commands we assign control valves. In the same figure, x_D and x_B represent the concentrations of the light component in the top product and in the bottom product.

The Relative Gain Array (RGA) [1] method is used to obtain an optimal pairing (i.e. minimum interactions between control loops) between the manipulated variables and controlled variables. Because the mass transfer (respectively the fractionation) is influenced by the liquid L and steam V fluxes that come into contact, it appears that for the quality control of the products extracted from the column, we must intervene upon the reflux flow rate L and/or the heating agent in the reboiler Q_{st}.

Reasons concerning material balance [2] impose the inclusion of the one of the manipulated variables: distilled (D) or residue (B) [2] among the manipulated

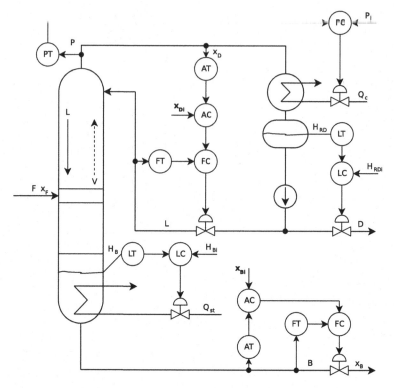

Fig. 7.6 L-B structure for dual concentration control of light component in *top* product and *bottom* product

variables associated to the control of the compositions. There are several possible pairs, in the Fig. 7.6 being represented the structure of control based on the pair *L-B*.

This structure uses as manipulated variables for the control of the compositions x_D and x_B the reflux flow rate L respectively the residue flow rate B. This structure is of type with direct material balance because one of the product flow rate, respectively the residue flow rate, is used as command for the control of one of the compositions. The levels H_B and H_{RD} are controlled by means of the residue flow rate B and the heating agent flow rate in the reboiler Q_{st}.

All the control systems in the Fig. 7.6 are feedback systems. In the case of the control systems for the compositions (the AC controllers) the durations of the transient regime are determined by the dynamics of the mass transfer (of the order of hours). In order to avoid the very long duration of the transient regime, during which the distillation products compositions are varying, we can use the feed-forward control. In the Fig. 7.7 it is presented a structure of feedforward control, in which there are considered the disturbances represented by the feed flow rate F and the concentration of the light component in the feed x_F.

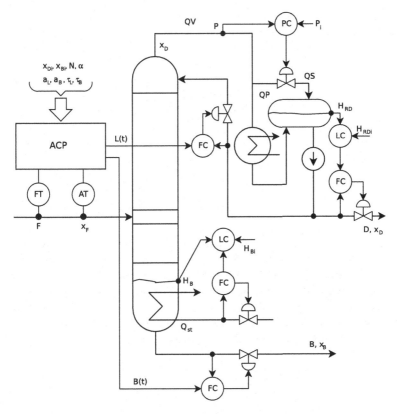

Fig. 7.7 Hierarchical control system of a fractionating column *ACP* dual feedforward concentration controller

The system illustrated in the Fig. 7.7 allows the determination and the application to the process of values of the manipulated variables L and B, so that the influence of the disturbances F and x_F upon the compositions x_D and x_B should be rejected. The stationary value of the command B is obtained from the relations associated to general balance and component balance,

$$F = B_{st} + D_{st} \tag{7.1}$$

$$Fx_F = B_{st}x_B + D_{st}x_D \tag{7.2}$$

Solving the system formed of the Eqs. (7.1) and (7.2) it results:

$$B_{st} = F\frac{x_D - x_F}{x_D - x_B} \tag{7.3}$$

The output value L_{st} can be determined using a simplified rapid design method adapted for control. One of these models is Douglas-Jafarey-McAvoy [3], that is based on a double expression of the separation factor S respectively:

$$S = \left(\frac{\alpha_m}{\sqrt{1 + 1/(Rx_F)}}\right)^N \tag{7.4}$$

$$S = \frac{x_D/x_B}{(1 - x_D)/(1 - x_B)} \tag{7.5}$$

$$\left(\frac{\alpha_m}{\sqrt{1 + 1/(Rx_F)}}\right)^N = \frac{x_D/x_B}{(1 - x_D)/(1 - x_B)} \tag{7.6}$$

In the relations (7.4) and (7.5), α_m is the mean relative volatility of the light component in comparison, N the number of theoretical trays and R the reflux ratio. From the relation (7.6) we determine R, and implicitly L_{st} considering that the reflux ratio is defined as the ratio between the outputs L and D ($R = L/D$).

In order that the effect of the disturbances should be synchronized with the one of the manipulated variables, the stationary model is completed with first order, deadtime elements, which ensure a delay of the commands application, respectively:

$$a_B \frac{dB(t)}{dt} + B(t) = B_{st}(t - \tau_B) \tag{7.7}$$

$$a_L \frac{dL(t)}{dt} + L(t) = L_{st}(t - \tau_L) \tag{7.8}$$

In the differential equations (7.7) and (7.8), the time constants a_L and a_B as well as the deadtimes τ_L and τ_B are determined by respecting the delays caused by the hydraulic phenomena from the column. By solving these equations in real time we obtain the dynamic values of the commands $L(t)$ and $B(t)$ which are applied as set points to the associated controllers.

The stationary model of control represented by the Eqs. (7.1)–(7.6) is valid only in the proximity of a mean functioning point of the process. When this point changes, the model must be adapted, variables used in adaptation being the relative volatility α and the number of theoretical trays N.

The structure of feedforward control, concisely presented above, has been successfully implemented in industry for a propylene-propane separation column [4, 5]. The results of the implementation have been quantified in the growth of conform propylene production correlated to the decrease of the separation effort, respectively of the flow rate of the steam in the reboiler of the column.

A parameter that directly influences the fractionation is the pressure. The processes of fractionation are designed taking into account a certain functioning pressure, for which reason this parameter must be maintained at a precise reference value. Usually the control of the pressure is achieved by intervening upon the quantity of thermal energy extract from the column. Because the biggest part of the thermal energy is given by the condensation of the steam, it appears that we can control the pressure by intervening upon the process of condensation.

In the case of the control structure presented in Fig. 7.5, the pressure is controlled by intervening upon the flow rate of the cooling agent Q_c. This solution is vulnerable due to the emergence, over a certain value of the flow rate Q_c of the phenomena of condensation saturation and implicitly of the possibility to control the pressure.

Much more efficient is the control of the pressure by intervening upon the condensation area from the steam space of the condenser. In the Fig. 7.7 it is presented this solution for the situation in which the condenser is situated under the reflux drum of the fractionation column. As it can be observed, the vapor flow Q_V is divided, the secondary vapor flow Q_S being used as command for the pressure control. We demonstrate [1] that the difference between the liquid levels from the reflux drum and the condenser can be controlled through the flow rate Q_E as manipulated variable. Considering that the level in the reflux drum is controlled, it appears that, by modifying the flow rate Q_E we can control the measure of the condensation area and implicitly the pressure.

7.1.3 Conventional Automatic Control of Heat Transfer Processes

Heat represents a form of energy specific to chemical processes. The processes that absorb heat are called endothermic and the processes that generate heat are called exothermic. Usually, thermal processes implies the production, exhaust and the transfer of heat. Taking into account the importance of heat exchange for chemical processes, we can admit that thermal processes have a strong interaction with chemical processes. As part of the technological equipment used by the thermal processes we can enumerate heating furnaces, steam generators, reboilers, condensers, etc.

No matter the type of thermic process, the quality objective depends on the amount of produced heat, exhausted or exchanged. From the control point of view, the most important parameter of a thermal process is represented by the temperature. Regarding efficiency, it is quantified by indicators like combustion efficiency, heat recovery rate, etc. Security objectives relate avoiding of environment pollution by dangerous emissions, explosions avoiding, etc.

The control function corresponding to these processes relates especially to temperature. In the following it will be presented as examples two control structures for heating furnaces with gas fuel. In such a furnace, the combustion process represented by the exothermic reaction between a fuel and an oxidant develops, the two components being also actuating quantities.

Fig. 7.8 Temperature and air/fuel ratio feedback control structure for a heating furnace: *RB* Ratio Block; K_i reference ratio

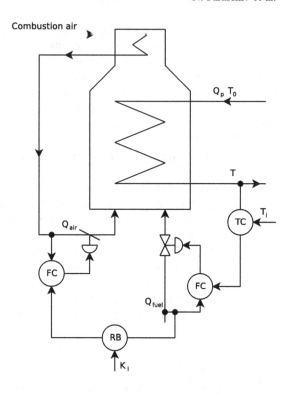

As the first control structure it is presented Fig. 7.8 in which the temperature is controlled in a feedback manner. As it can be observed, the temperature is controlled with the aid of combustion flow Q_{fuel}. The combustion is controlled by adjusting air flow with the aid of the ratio block RB, as a function of fuel flow. Closed loop control of temperature presents the disadvantage of a non-steady state in the case of set point change or disturbances.

The second variant, presented in Fig. 7.9 obtains temperature control by a feedforward structure, taking into account the disturbances represented by the feed flow Q_p and its temperature T_0.

The algorithm associated with the controller TC is, in this case, process dependent and reflects the heat transferred to the product heated by burning the fuel as expressed in the following thermal balance equation:

$$Q_{fuel}q_{fuel} = Q_p c_p (T_i - T_0) + W_1 \tag{7.9}$$

where:

- Q_{fuel}—fuel flow rate
- q_{fuel}—fuel thermal value
- Q_p—product flow rate
- c_p—product specific heat capacity

Fig. 7.9 Temperature and air/fuel ratio feedforward control structure for a heating furnace: *RB* Ratio Block; K_i reference ratio

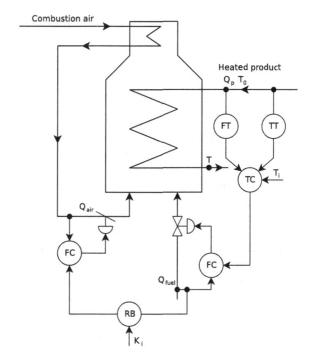

- T_i—temperature reference value
- T_0—product temperature at the furnace input
- W_1—flow rate of heat losses

From Eq. (7.9) it can be obtained the stationary state fuel flow rate:

$$Q_{\text{fuel\_st}} = Q_p \frac{c_p}{q_{\text{fuel}}}(T_i - T_0) + \frac{W_1}{q_{\text{fuel}}} \tag{7.10}$$

where: Q_p and T_0 are measured and T_i, c_p, q_{fuel} and W_1 are considered known constants.

In order to obtain the dynamic regime output it is necessary to solve the following differential equation associated with the dynamics of heat transfer process:

$$a_T \frac{dQ_{\text{fuel}}(t)}{dt} + Q_{\text{fuel}}(t) = Q_{\text{fuel\_st}}(t - \tau_T) \tag{7.11}$$

The dynamic section of the model, represented by (7.11) is justified by the necessity of applying the output value, that means the changing of fuel flow rate respectively in accordance to the heat transfer process dynamic behavior.

7.1.4 Conventional Automatic Control of Chemical Reactors

The control of a chemical reaction poses problems regarding the stoichiometry, thermodynamics and kinetics of the reaction. Deciding the appropriate control structure depends also on the type of reactor in which the chemical reaction takes place.

Stoichiometry of the chemical reaction allows fixing the ratios between the amounts of reactants that make the reaction possible. From the automatic control point of view, it is necessary to provide a certain ratio between the reactants flow rate.

Applying thermodynamics tools to a chemical reaction allows the evaluation of the reaction heat and of the equilibrium conversion rate, respectively [1]. The development of a chemical reaction at equilibrium requires meeting certain values for temperature and pressure, automatic control of these parameters respectively.

Chemical reaction kinetics studies mainly the reaction rate of the reaction development. When maintaining a certain reaction rate, temperature control is essential. The existence of catalysts provides, in certain conditions, an increase in the reaction rate. From the point of view of automatic control, the ratio of reactant flow rate and of catalyst flow rate can represent a process action for the reaction rate automatic control.

As far as the chemical reaction equipment host is concerned, chemical reactors respectively, there are diverse types of reactor from which we mention the Continuous Stirred-Tank Reactors and the Tubular Reactors.

From the automatic control point of view, continuous stirred-tank reactors are treated as concentrated-parameter systems. In other words, inside the reactor, parameters values (temperature, composition of reactants, conversion rate, pressure, etc.) are only functions of time and not of spatial coordinates.

As shown in Fig. 7.10, in order to control the temperature inside a continuous stirred-tank reactor there should be varied the heat carrier flow by means of a cascade temperature-temperature of the heat carrier control system. Cascade control provides the advantage of compensating all the disturbance effects that affect heat carrier temperature. The reactor load is also controlled when using the $A1$ reactant flow rate as a control action and the reactor's holdup, actuating upon the reactor's output flow rate.

For the reactors for which the number of moles of substance is changing, such as the polymerization reactions, it is necessary to control the pressure, such an example being presented in Fig. 7.11.

In order to control the pressure, the reaction product flow rate is used as a control value. Also, the flow rates of the reactants are controlled in order to maintain a proper development of the reaction.

Tubular reactors are distributed parameter systems characterized by variable parameters, as functions of both time and spatial coordinates. Specific to such a reactor are the following parameters: temperature, composition, conversion rate, etc.

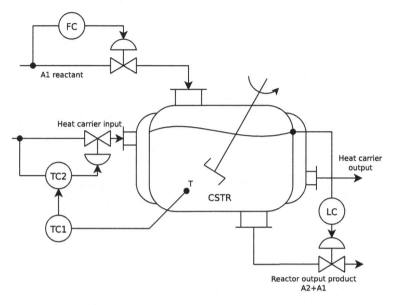

Fig. 7.10 Control structure of a Continous Stirred-Tank Reactor

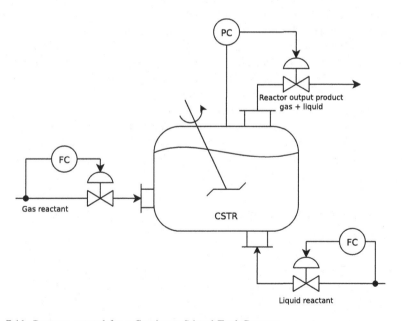

Fig. 7.11 Pressure control for a Continous Stirred-Tank Reactor

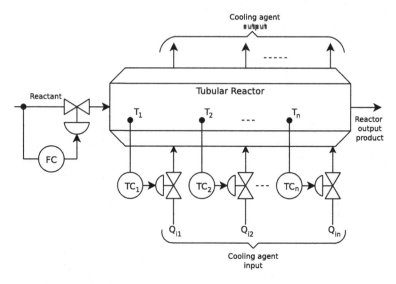

Fig. 7.12 Temperature profile control along a tubular reactor

For such a reactor, the main objective of the automatic control is represented by the fulfillment of quality specifications of the reaction product. Among these specifications, a special parameter is considered the composition of the product. Because composition measurement poses several difficulties, including high inertia, it is preferred the control of the reaction with the aid of temperature. Practically, it is necessary to control the temperature profile along the reactor, the reactor itself being a composition profile indicator.

For these reactors, in which strong exothermic reactions take place, it is necessary to consider the injection of the cooling medium in many points.

It can be observed in Fig. 7.12 that every injected flow rate is used as an action control for controlling the temperatures from different points of the tubular reactor.

7.2 Computational Intelligence Techniques for Process Control

The most advanced process control strategies are model-based and make use of some artificial intelligence techniques, usually applied to non-algorithmic problem solving, such as expert systems, artificial neural networks, genetic algorithms etc. Computational intelligence was introduced in 1994 [6] as a paradigm that combines three main complementary computing technologies: fuzzy computing (based on fuzzy logic and fuzzy sets), neural computing (based on artificial neural networks) and evolutionary computing (based on genetic algorithms and evolutionary strategies). Recent developments in this area revealed the efficiency of using the new computational intelligence techniques such as those provided by swarm intelligence (e.g. particle swarm optimization, ant colony optimization).

In literature is presented a set of applications of artificial intelligence techniques in various processes control, such as BIOEXPERT, AQUALOGIC, etc. [7–10]. In this subchapter it is presented an introduction to the basic computational intelligence techniques (fuzzy systems, artificial neural networks and genetic algorithms) and a brief overview of some computational intelligence applications in chemical process control.

7.2.1 Computational Intelligence Techniques

The main basic computational intelligence techniques are fuzzy systems, artificial neural networks and genetic algorithms. Fuzzy systems are a proper technique for imprecision and approximate reasoning, artificial neural networks for learning, and genetic algorithms for optimization.

Fuzzy systems combines fuzzy logic with fuzzy sets theory [11], the key idea being that truth values (from the fuzzy logic) and the membership values (from the fuzzy sets theory) are real values in the interval [0, 1], where 0 means absolute false, and 1 absolute true. A fuzzy system is developed by structuring the domain knowledge (provided by the human experts from the chosen field of application) under the form of linguistic variables and fuzzy rules set. A fuzzy set can be defined by assigning to each possible object a value that represent the fuzzy set membership degree. The fuzzy sets theory express imprecision quantitatively by introducing the membership degree from it is not member to it is totally member. If F is a fuzzy set than the μ membership function measures the degree under which x is a member of F. This membership degree represents the possibility that x can be described by F. Each membership function, specific to a certain fuzzy term, is represented by four parameters grouped in the T_i term: $T_i = (a_i, b_i, c_i, d_i)$, corresponding to the weighted interval from Fig. 7.13. The form of the membership function can be triangular, trapezoidal, Gaussian, sigmoidal etc., depending on the application.

In the sets theory the *high*, *medium* and *small* symbolic values of the temperature variable, for example, have mutual exclusive associated values. If the numerical value of the temperature is smaller than 100 °C then the symbolic value is small, if the value is in the interval [100 °C, 300 °C] then the value is *medium*, and if it is greater than 300 °C then the value is *high*, as shown in Fig. 7.14, where there are overlappings in the neighbourhood of the interval limits (100 and 300 °C).

Figure 7.15 presents the general scheme of a fuzzy system. The role of the fuzzy system is to make fuzzy inferences that interpret the input values and based on a fuzzy rules set assign values to the outputs.

The most important engineering applications of fuzzy systems are control applications (e.g. system control and process control). Figure 7.16 shows the block

Fig. 7.13 The membership function parametric representation

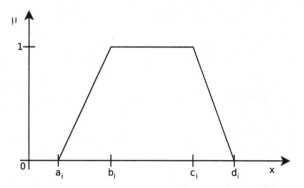

Fig. 7.14 Examples of fuzzy values for the temperature variable

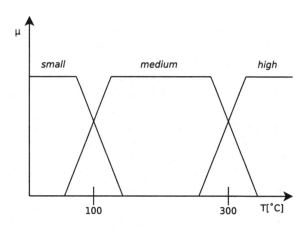

Fig. 7.15 The general scheme of a fuzzy system

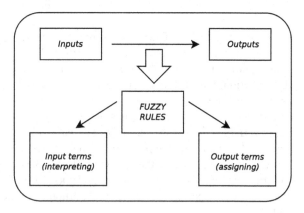

diagram of a fuzzy control system. A fuzzy inference system (FIS) is a non linear system that applies **if-then** fuzzy rules and can model the qualitative aspects of the human knowledge and of the reasoning processes without accurate quantitative analysis. The fuzzy logic modeling techniques can be classified in three categories:

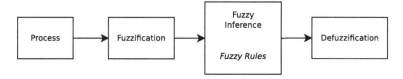

Fig. 7.16 The block diagram of a fuzzy control system

the Mamdani type linguistic technique, the relational equation, and the Takagi-Sugeno-Kang (TSK) technique.

Artificial neural networks are universal approximators [12], capable to learn complex mappings that are dependent on their structures. An artificial neural network (ANN) is composed by a number of processing units named neurons that are connected under a specific topological structure. Each connection has associated a numerical weight, that is usually, randomly initialized and later determined, during the neural network training process. Each neuron has an activation function that allow information transmission toward other neurons. The construction of a neural network for solving a certain problem consists in setting the network topology with the number of layers, number of neurons (for each layer: input, output, hidden), the type of each neuron (i.e. activation function) and the way of interconnecting the neurons. The next step is weights initialization and the network training by applying a training algorithm to a training set for which the final values of the weights are computed.

In Fig. 7.17 it is given the block diagram of a generic artificial neural network.

The generic neural computing algorithm written in pseudocode is given in Fig. 7.18.

The first step of the algorithm makes weights initialization, usually with random values from the interval [0, 1]. During the second step the network is trained by using a training algorithm (one of the most used is backpropagation and its various improved versions) and a training set, and the final values of the weights are determined. In the last step, the artificial neural network is tested and validated on specific testing and validation data sets.

The main neural network topological structures are feedforward and recurrent. Examples of neural networks types are feedforward neural networks, radial based neural networks, Elman neural networks, Hopfield neural networks, Kohonen neural networks, Boltzman neural networks, probabilistic neural networks, etc.

Artificial neural networks are applied with success in various applications: pattern recognition, time series prediction, optimization problems (such as optimal control) etc. They are a proper tool for solving engineering problems that have complex noisy input data.

Genetic algorithms (GA) are a subclass of evolutive algorithms that are optimization methods that mimics the processes that appear in genetics and natural evolution [13]. They maintain a population of solutions (named individuals) that evolve in time by applying genetic operators of selection, recombination and

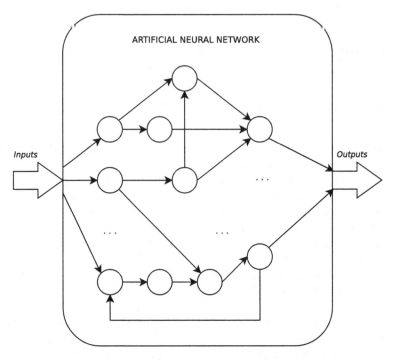

Fig. 7.17 The block diagram of a generic ANN

Fig. 7.18 The generic ANN computing algorithm

```
Algorithm ANN_Technique(ANN, Inputs, Outputs)
// W – the ANN weights matrix for each layer
// Training_set – set of known (Inputs, Outputs) pairs
{
    1. ANN_weights_initialization(ANN, W);
    2. ANN_training(ANN, W, Training_set);
    3. ANN_testing_validation(ANN, W, Inputs, Outputs);
}
```

mutation. The rate of applying mutation operators is much smaller than the recombination rate.

A genetic algorithm provides an efficient optimization technique, being a stochastic algorithm. Several solutions are investigated in parallel. The final solution is the global optimal solution. A genetic algorithm is a search algorithm that finds the best solution that maximizes a fitness function (FF) that is problem dependent. In the classical variant of a genetic algorithm, a solution is represented as a string from a finite alphabet, each element of the string being a gene. In general, the string is a string of bits. The classical form of a genetic algorithm is provided in Fig. 7.19 under the form of a function written in pseudocode.

Genetic algorithms can be applied to process control, either as a standalone technique or in combination with other computational intelligence techniques.

Fig. 7.19 The general form
of a classical genetic
algorithm

```
Function GeneticAlgorithm (population, FF) return solution
Inputs:
        population; // a set of solutions
        FF;        /* the fitness function matching a solution */
{
        repeat
        parents      ←   Selection(population, FF);
        population   ←   RecombinationMutation(parents);
        until *      a solution has the best matches according to
                     the FF fitness function
        return *     the best solution according to the FF function;
}
```

The new computational intelligence techniques provided by swarm intelligence are part of the evolutionary computing. They are using nature inspired collective intelligence. Examples of such techniques are particle swarm intelligence (PSO) and ant colony optimization (ACO). These techniques can be applied also to process control.

As stated initially in [14] and lately confirmed by the research results reported in the literature, combinations of computational intelligence techniques (i.e. hybrid techniques) are more effective and robust in process control.

One of the most used hybrid fuzzy systems is ANFIS—the Adaptive Neuro-Fuzzy Inference System introduced in [15], that is a FIS based on adaptive artificial neural networks. ANFIS uses the TSK model and is actually a FIS implemented under the form of an artificial neural network. Each layer of the network corresponds to a part of the FIS and the FIS parameters are codified as the artificial neural network weights.

7.2.2 Applications of Computational Intelligence in Chemical Process Control

Several applications of computational intelligence in chemical process control were reported in the literature. Some of them are using one of the three main computational intelligence techniques (FIS, ANN, GA), while others are using combinations of these techniques. In this section we are making a brief presentation of selected applications, grouped by the computational intelligence technique that was applied.

7.2.2.1 Fuzzy Systems Applications

In [16] it is presented a fuzzy logic control used as a promising control technique for improved process control of a fluid catalytic cracking unit in refinery process industry. A recent example of applying real time fuzzy control to a *pH* neutralization process is given in [17]. The experiments were done at laboratory level and showed a good behavior of the proposed PI fuzzy controller. Another example of a *pH* fuzzy controller was proposed more than a decade ago in [18]. A fuzzy dynamic learning controller was proposed in [19] for time delayed, non linear and unstable chemical processes control.

7.2.2.2 Artificial Neural Networks Applications

In [20] it is tackled the modeling problem of complicated batch processes in the context of model-based control of chemical processes. The authors proposed a novel hybrid neural network, called a structure approaching hybrid neural network (SA-HNN), for intelligent modeling of a batch reactor with partially unmeasurable states. The predictive control of a wastewater treatment process is described in [21]. A predictive controller based on a feedforward artificial neural network as internal model of the process, alters the dilution rate and control the concentration of the dissolved oxygen. The artificial neural networks approach was used in the last two decades as a powerful tool for a wide range of applications in the oil and chemical industry. A recent example is reported in [22], where a feedforward neural network was applied to model the desalting and dehydration process, with the purpose of optimizing the whole chemical process and increasing the efficiency of oil production. In [23] it is presented a review of some applications of artificial neural networks in chemical process control, at simulation and online implementation level. Most of the reported applications use feedforward neural networks. As shown in [24] the most popular domain in which artificial neural networks were applied is chemical engineering. The applications reported by authors included chemical process control optimization. Finally, a computer simulation study of industrial process control of chemical reactions by using spectroscopic data and artificial neural networks is described in an older research work [25].

7.2.2.3 Genetic Algorithms Applications

In [26] it is presented an intelligent technique based on genetic algorithms for optimal controller tuning in a *pH* neutralization process. The experimental results showed the capability of the genetic algorithm to quickly adapt the controller to dynamic plant characteristic changes in the *pH* neutralization process. The biogas plant control and optimization by using genetic algorithms and particle swarm optimization is discussed in [27]. The authors apply the two computational intelligence techniques (GA and PSO) for the optimization of the substrate feed with regard to its flow rate and composition in the case of a biogas plant. The use of two computational intelligence techniques, genetic algorithms and fuzzy systems, for fed-batch fermentation process control is presented in [28]. The experimental results showed that the two techniques performed better than conventional optimizations methods in the presence of noise, parameter variation and randomness.

7.2.2.4 Swarm Intelligence Applications

In [29] it is discussed in detail the implementation of particle swarm optimization (PSO) algorithm in PID tuning for a controller of a real time chemical process. Particle swarm optimization is an evolutionary computation algorithm that simulates social behavior in swarms (e.g. bird flocking and fish schooling). The main advantages of the proposed swarm intelligence based solution are given by its simplicity and low cost, as well as by its good performance in case of PID controllers tuning. Another application of using PSO is described in [27]. The application of another swarm intelligence technique, ant colony optimization (ACO) is reported in [30]. The authors proposed a new optimal method for designing and computing the parameters of an ACO-based controller for non linear systems described by TSK models (Fig. 7.19).

7.2.2.5 Hybrid Computational Intelligence Applications

The neuro-fuzzy control of chemical technological processes is discussed in [31]. A combination of the predictive and ANFIS controller was proposed and tested as intelligent control system for a Continuous Stirred-Tank Reactor (CSTR) control problem. The experiments showed better results than those obtained with the original predictive and PID controller. Another successful application of neuro-fuzzy intelligent process control is presented in [32].

7.3 Case study: The Wastewater pH Neutralisation Process in a Wastewater Treatment Plant

Through treatment process it is understood the set of physical procedures (that compose the physical wastewater treatment plant (WWTP) step), physical-chemical procedures (physical-chemical WWTP step) and the biological ones (the biological step) through which is achieved the pollutants removing from wastewater. Such procedures are: neutralization, flotation, absorption, extraction, etc. The pH neutralization process is achieved in the WWTP physical-chemical step, in chemical reactors of high capacity, the so called Continuous Stirred Tank Reactors (CSTR). The quality indicator pH is a measure of solution acid or alkaline (basic) character and is measured on a scale from 0 to 14 pH units. For acid type wastewater neutralization (with $pH < 7$), namely for increasing the pH value, are used basic (alkaline) type substances, such as: lime (calcium lime-CaO) under calcium hydroxide form-$Ca(OH)_2$), dolomite (calcium and magnesium carbonate), limestone, hydroxide sodium ($NaOH$), etc. For alkaline (basic) type wastewater neutralization (with $pH > 7$), namely for decreasing the pH value, are used: sulphuric acid (H_2SO_4), carbonic acid (H_2CO_3) and chlorine hydride (HCl) [33]. In Fig. 7.20 are presented examples of dynamic characteristics (the process response in time) for pH neutralization process [1].

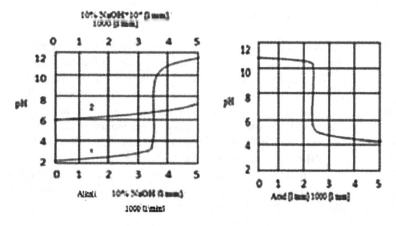

Fig. 7.20 Dynamic characteristics for acid and alkaline *pH* neutralization

As it can be observed in Fig. 7.20, the wastewater *pH* neutralization process is a complex one having a high nonlinear behaviour.

In this case study is presented a neuro-fuzzy system developed for wastewater *pH* control, system that has an ANFIS (Adaptive Neuro-Fuzzy Inference System) controller. Also, it was developed a mathematical model (under a transfer function form) of the wastewater *pH* neutralization process. The architecture of the proposed automatic control system for wastewater *pH* control (pHACS) is presented in Fig. 7.21.

As it can be observed in Fig. 7.21, the pHACS components are [34]:

1. The process represented by the developed mathematical model (transfer function) for wastewater *pH* neutralization process
2. The controller (R-ANFIS) developed using neuro-fuzzy systems (ANFIS); for controller development were used the facilities offered by Matlab 7.9 environment through ANFIS Editor GUI
3. Two actuators EE1 and EE2; EE1 is the acid-type neutralizer (H_2SO_4) dosing pump while EE2 is the alkali-type neutralizer (*NaOH*) dosing pump; the functioning of one of these two pumps depends on the *pH* character (acid or alkaline)
4. A *pH* meter for measuring the controlled variable (*pH*) value at the process output
5. The wastewater *pH* set point (r_{pH}), value established at national level through a special normative in domain, called NTPA-001/2002 [35]
6. The command (*c*) generated by the system controller (R-ANFIS), controller that, depending on the error value, generates the command for the necessary neutralizer agent flow for bringing the controlled variable (*pH*) to its set point
7. The error (*e*) defined as the difference between the *pH* set point (r_{pH}) and the measured *pH* value at the process output (m_{pH})

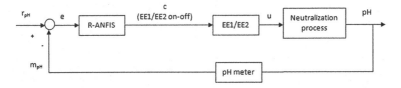

Fig. 7.21 pHACS architecture [34]

8. The manipulated variable (u) that represents neutralizer agent flow dosage
9. *pH* is the controlled output
10. m_{pH} is the measurement signal

Hereinafter is presented the development of the wastewater *pH* neutralization process mathematical model (under a transfer function form) and the development of the R-ANFIS controller as a component of the proposed pHACS.

7.3.1 The Process Mathematical Model Development

The analyzed process is that of wastewater *pH* neutralization, process that takes place in a wastewater treatment plant (WWTP) physico-chemical step. For this type of process, studying the literature was chosen the mathematical model developed by Ibrahim R. in his PhD Thesis [36]. The mathematical model of the process was first of all analyzed and then implemented (simulated) in Matlab 7.9/ Simulink environment in [34]. According to [36] and [1], the *pH* neutralization process has a high-nonlinear behaviour. Due to the process model complexity, in order to obtain that transfer function (the simplified mathematical model of the process) that better describes the process was applied the model linearization. The *pH* neutralization process inputs and outputs are presented in Fig. 7.22 [34].

As it can be observed in Fig. 7.22, the process inputs and outputs are:

1. F_1 is the acid stream flow rate;
2. F_2 alkaline stream flow rate;
3. C_1 is the acid concentration in basin;
4. C_2 is the alkalinity concentration in basin;
5. $y(pH)$ is the process output.

For model linearization (for obtaining the transfer function) we need to calculate the proportional control factor (K_p) and the process transient time (T_p), defined as follows:

$$K_p = \frac{\Delta y}{\Delta F_1} \tag{7.12}$$

$$K_p = \frac{\Delta y}{\Delta F_2} \tag{7.13}$$

Fig. 7.22 Process inputs/
outputs [34]

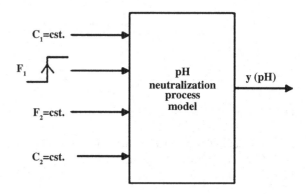

$$T_p = \frac{T_{tr}}{4} \tag{7.14}$$

Δy is the output variation (pH variation), ΔF_1 is the input F_1 variation (acid type-neutralizer flow variation), ΔF_2 is the input F_2 variation (alkali type-neutralizer flow variation) and T_{tr} is the transient regime duration.

As it can be observed in Fig. 7.22, we considered F_1 (acid type neutralizer agent) step input in the process and the others inputs were maintained constant. In this case the process dynamic response is presented in Fig. 7.23.

Using (7.12) and knowing according to [36] the domains for F_1 ($F_1 \in [0.0.260]l/h$) and for pH ($pH \in [0.0.14]$ $pHunits$), we have:

1. $K_p = -0.72$;
2. $T_{tr} = t(y(0.98x\Delta y)) = 0.795$;
3. $T_p = 0.19875$;
4. $G_{yF_1} = \frac{K_p}{T_p s + 1} = -\frac{0.72}{0.19875s + 1}$ (transfer function)

Using the same reasoning was also achieved the linearization for F_2 domain ($F2 \in [0.0.340]l/h$), as it can be observed in Fig. 7.24.

We considered F_2 (alkali-type neutralizer agent) step input in process and the others inputs are maintained constant. The process response at input step F_2 is that presented in Fig. 7.25.

Using (7.13) and (7.14) were obtained:

1. $K_p = 1.72$
2. $T_p = 0.19875$
3. $G_{yF_2} = \frac{K_p}{T_p s + 1} = \frac{0.72}{0.19875s + 1}$ (transfer function)

Using the linearization of the model, we obtained the searched transfer function that will be used as the model of the wastewater pH neutralization process.

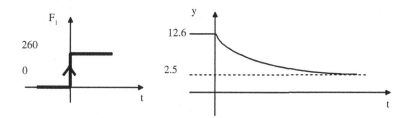

Fig. 7.23 The process response at step input F_1 [34]

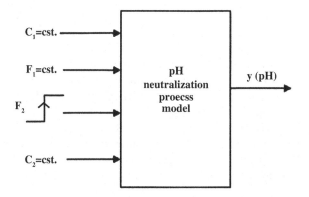

Fig. 7.24 Process inputs/outputs [34]

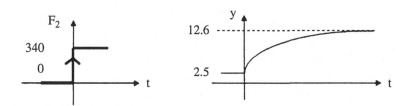

Fig. 7.25 The process response at step input F_2 [34]

7.3.2 The R-ANFIS Controller Development

As we have mentioned for developing the R-ANFIS controller from Fig. 7.21, were used the facilities supplied by Matlab 7.9, through the *anfisedit* command usage, command that calls the ANFIS Editor.

In 26, R-ANFIS (RpH2) is considered to be a first-order Sugeno type fuzzy system with one input (error) and one output (EE1/EE2 opening degree for acid or alkaline/basic neutralizer agent dosage).

Table 7.1 R-ANFIS rule base

No.		Error		EE1/EE2 open degree
1		in1mf1		out1mf1
2		in1mf2		out1mf2
3		in1mf3		out1mf3
4		in1mf4		out1mf4
5	If	in1mf5	Then	out1mf5
6		in1mf6		out1mf6
7		in1mf7		out1mf7
8		in1mf8		out1mf8
9		in1mf9		out1mf9
10		in1mf10		out1mf10

The rules base contains a number of ten rules, as it can be observed in Table 7.1, rules automated generated through the usage of *Generate Fis* option, option that based on the training data (data obtained through the process analysis) from Table 7.2, generated the system with fuzzy inference (FIS) (FIS with the structure presented in Fig. 7.26).

In Table 7.1, ERROR represents the controller input, defined as the difference between the *pH* set point and the *pH* measurement at the process output, while EE1/EE2 OPEN DEGREE is the controller command (output), defined to be the EE1 or EE2 opening degree for acid or alkali type neutralizer agent flow necessary for *pH* control.

After the automatically obtaining of FIS (Fig. 7.26), can be visualized the generated ANFIS model using *Structure button* from user graphical interface (GUI). In Fig. 7.27 is presented the ANFIS model structure (Table 7.2).

In Fig. 7.27 we have a model with one input (error), one output (EE1/EE2 opening degree for acid or alkali type neutralizer agent dosage) and also a number of ten fuzzy rules.

As it can be observed in Fig. 7.28, for training the generated fuzzy inference system (FIS), was used a hybrid training algorithm, that according to [37] has two steps: feed forward-propagation and back-propagation.

For model validation was used a validation data set. As it can be observed in Fig. 7.29, we can say that the generated model is a valid one (validation data output follows the FIS output).

In Fig. 7.30 is presented the application *Rule Viewer* that shows a map of the entire fuzzy inference process.

In Fig. 7.31 is presented under a graphical form the relation between the R-ANFIS controller input (*error*) and output (command-*EE1/EE2 OPEN DEGREE*).

Having developed the R-ANFIS controller it can be developed the automated system for wastewater *pH* control (*pHACS*) in Simulink.

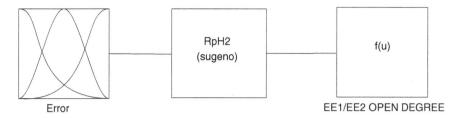

Fig. 7.26 R-ANFIS architecture

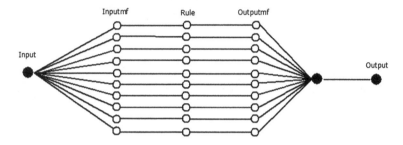

Fig. 7.27 ANFIS model structure

Table 7.2 Training data

Error	EE1/EE2 open degree
5	100
4	75
3	50
2	37
1	25
0	0
-1	-25
-2	-37
-3	-50
-4	-75
-5	-100

7.3.3 pHACS Implementing in Matlab/Simulink

Using the fuzzy inference system presented in Fig. 7.26, generated and trained with the help of an artificial neural network (ANN), was developed the R-ANFIS controller. For implementing this controller in Simulink was used *Fuzzy Logic Controller* with *Ruleviewer block*, as it can be observed in Fig. 7.32.

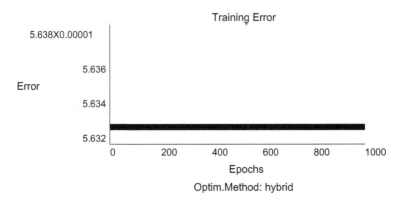

Fig. 7.28 FIS training

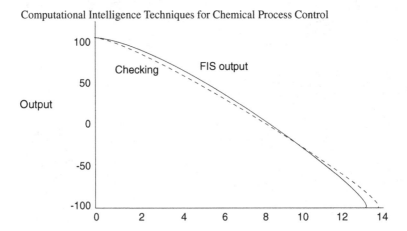

Fig. 7.29 ANFIS model validation

In Fig. 7.33 is presented the neuro-fuzzy automatic system (pHACS) for an alkali type *pH* control using the transfer function G_{yF_1} obtained through model linearization (3.1).

The pHACS for alkali-type *pH* control response is presented in Fig. 7.34.

In Fig. 7.35 is presented the pHACS architecture for acid-type *pH* neutralization using the transfer function G_{yF_2}.

The pHACS for acid-type *pH* control response is presented in Fig. 7.36.

The experimental results obtained for the above mentioned experiments are presented in Table 7.3.

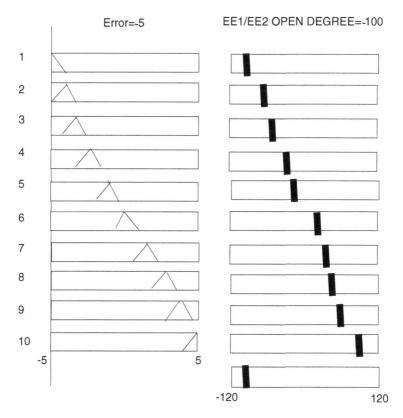

Fig. 7.30 R-ANFIS *Rule Viewer*

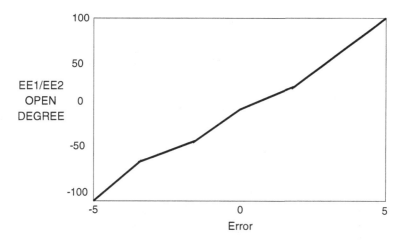

Fig. 7.31 R-ANFIS controller

Fuzzy Logic Controller with Ruleviewer
FIS with a ruleviewer for fuzzy logic rules
Parameters
FIS matrix
RpH2
Refresh rate (s)
2

Fig. 7.32 Fuzzy Logic Controller with Ruleviewer block

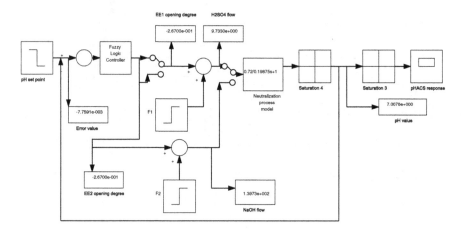

Fig. 7.33 pHACS for alkali-type *pH* control using G_{yF_1} (Experiment no.1)

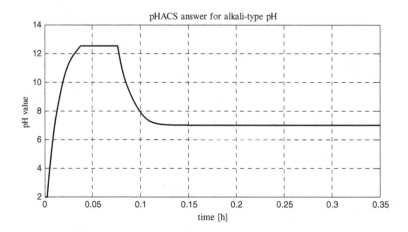

Fig. 7.34 pHACS response for alkali-type *pH* (Experiment no.1)

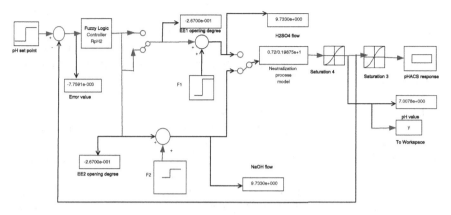

Fig. 7.35 pHACS for acid-type *pH* control using G_{yF_2} (Experiment no.2)

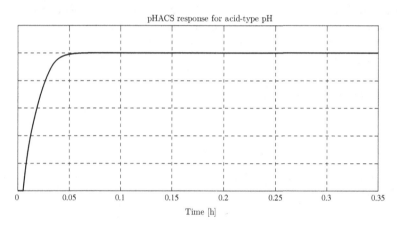

Fig. 7.36 pHACS response for acid-type *pH* (Experiment no.2)

Table 7.3 Experimental results

No.exp.	$F_1H_2SO_4$ (l/h)	F_2NaOH (l/h)	*pH* set point	*pH* process value	pHACS error $(e = i_{pH} - m_{pH})$	Transient regime duration T_{tr}(h)
1	10	140	7	7.0078	0.0077591	0.35
2	10	10	7	7.0078	0.0077591	0.35

7.4 Conclusion

As it can be observed in Table 7.3 using neuro-fuzzy control, the *pH* was brought very close to its set point (7.7), therefore the controller (R-ANFIS) obtained using neuro-fuzzy techniques and the developed automatic control system (pHACS) are supplying good results, such as low error.

Through the usage of artificial neural networks (ANN), especially to theirs capacity to learn and to adapt, the fuzzy systems (FIS) performances are considerably improved. So, the parameters of a fuzzy system (FIS set of rules and membership functions) are calculated through learning (training) methods using input-output data sets.

The usage of neuro-fuzzy controllers obtained through the development, training and testing of a Sugeno type fuzzy system can be a viable solution for processes with essential nonlinearities, as in case of the wastewater pH neutralization process.

The applicability of artificial intelligence (AI) techniques (fuzzy logic, artificial neural networks, neuro-fuzzy systems, expert systems, etc.) in control problems is justified due to the AI techniques advantages (for instance, fuzzy logic is indicated to be used for complex and nonlinear process control, etc.) that AI brings to the control domain.

Computational intelligence provides low cost robust solutions with a good tolerance of imprecision and uncertainty, being a proper tool for process control in real time, where the tradeoff is between accuracy and processing speed. The most efficient chemical process control methods are those based on hybrid approaches that combine different computational intelligence techniques.

References

1. Marinoiu, V., Paraschiv, N.: Automatizarea Proceselor Chimice. Editura Tehnica Publishing House, Bucuresti (1992)
2. Shinskey, G.F.: Distillation Control for Productivity and Energy Conservation. McGraw-Hill Book Company, New York (1984)
3. Jafarey, A., Douglas, M.J., Mc Avoy, J.T.: Short-cut techniques for distillation column design and control -1 column design. Ind. Eng. Chem. Process Dev **18**, 2 (1979)
4. Paraschiv, N., Cirtoaje, V.: Sistem automat evoluat pentru procesul de separare a propenei de chimizare—Implementarea industriala. Rev. Chim. **43**(7), 390–397 (1992)
5. Paraschiv, N.: Echipamente si programe de conducere optimala a proceselor de fractionare a produselor petroliere. Ph.D. Thesis, Petroleum-Gas Institute (1987)
6. Bezdek, J.C.: What is computational intelligence? In: Zurada, J.M., Marks II, R.J., Robinson, C.J. (eds.) Computational Intelligence—Imitating Life, IEEE World Congress on Computational Intelligence—WCCI, pp. 1–12. IEEE Computer Society Press, Piscataway (1994)
7. Cărbureanu, M.: A system with fuzzy logic for analyzing the emissary Pollution level of a wastewater treatment plant. In: Proceedings of the 18th International Conference on Control Systems and Computer Science—CSCS-18, Bucharest, Politehnica Press (2011), pp. 413–420
8. Lapointe, J., et al.: BIOEXPERT-an expert system for wastewater treatment process diagnosis. Comput. Chem. Eng **13**, 619–630 (1989)
9. Robescu, D., et al.: The Automatic Control of Wastewater Treatment Processes, pp. 339–347. Technical Press, Bucharest (2008)
10. Passavant Geiger (2012). http://www.masons.co.nz/sites/default/files/imce/Products%2B Service_engl.pdf, Cited 5 April 2012
11. Zadeh, L.A.: Fuzzy sets. Inf. Control **8**, 338–353 (1965)

12. Rumelhart, D., Widrow, B., Leht, M.: The basic ideas in neural networks. Commun. ACM **37**, 87–92 (1994)
13. Goldberg, D.E.: Genetic Algorithms in Search, Optimization and Machine Learning. Addison-Wesley, Massachusetts (1989)
14. Zadeh, L.A.: Fuzzy logic, neural networks and soft computing. Commun. ACM **37**, 939–945 (1994)
15. Jang, J.S.R.: ANFIS: adaptive-network-based fuzzy inference system. IEEE Trans. Syst. Man. Cybern. **23**, 665–685 (1993)
16. Taskin, H., Cemalettin, K., Uygun, O., Arslankaya, S.: Fuzzy logic control of a fluid catalytic cracking unit (FCCU) to improve dynamic performance. Comput. Chem. Eng. **30**, 850–863 (2006)
17. Fuente, M.J., Robles, C., Casado, O., Syafiie, S., Tadeo, F.: Fuzzy control of a neutralization process. Eng. Appl. Artif. Intell. **19**, 905–914 (2006)
18. Benz, R., Menzl, S., Stühler, M.: A self adaptive computer based pH measurement and fuzzy control system. Wat Res. **30**, 981–991 (1996)
19. Song, J.J., Park, S.: A fuzzy dynamic learning controller for chemical process control. Fuzzy Sets Syst. **54**, 121–133 (1993)
20. Cao, L.L., Li, X.G., Jiang, P., Wang, J.: Intelligent modelling of a batch reactor with partially unmeasurable states based upon a structure approaching hybrid neural networks. J. Syst. Control **223**, 161–173 (2009)
21. Caraman, S., Sbârciog, M., Barbu, M.: Predictive control of wastewater treatment process. Int. J. Comput. Commun. Control **II**, 132–142 (2007)
22. Al-Otaibi, M.B., Elkamel, A., Nassehi, V., Abdul-Wahab, S.A.: A computational intelligence based approach for the analysis and optimization of a crude oil desalting and dehydration process. Energy Fuels **19**, 2526–2534 (2005)
23. Hussain, M.A.: Review of the application of neural networks in chemical process control, simulation and online implementation. Artif. Intell. Eng. **13**, 55–68 (1999)
24. Baughman, D.R., Liu, J.A.: Neural Networks in Bioprocessing and Chemical Engineering. Academic Press, San Diego (1995)
25. Puebla, C.: Industrial process control of chemical reactions using spectroscopic data and neural networks: a computer simulation study. Chemometr. Intell. Lab. Syst. **26**, 27–35 (1994)
26. Valarmathi, K., Devaraj, D., Radhakrishnan, T.K.: Intelligent techniques for system identification and controller tuning in pH process. Braz. J. Chem. Eng. **26**, 99–111 (2009)
27. Wolf, C., McLoone, S., Bongards, M.: Biogas plant control and optimization using computational intelligence methods. Automatisierungstechnik **57**, 638–649 (2009)
28. Riid, A., Rustern, E.: Computational intelligence methods for process control: fed-batch fermentation application. Int. J. Comput. Intell. Bioinf. Syst. Biol. (2009). doi:10.1504/IJCIBSB.2009.030646
29. Giriraj Kumar, S.M., Sivasankar, R., Radhakrishnan, T.K., Dharmalingam, V., Anantharaman, N.: Particle swarm optimization technique based design of Pi controller for a real-time non-linear process. Instrum. Sci. Technol. 36:525–542 (2008)
30. Liouane, H., Douik, A., Messaoud, H.: Design of optimized state feedback controller using ACO control law for nonlinear systems described by TSK models. Stud. Inform. Control **16**, 307–320 (2007)
31. Blahová, L., Dvoran, J.: Neuro-fuzzy control of chemical technological processes. In: Fikar, M., Kvasnica, M. (eds.) Proceedings of the 17th International Conference on Process Control, pp. 268–272. Slovakia (2009)
32. Chen, C.T., Peng, S.T.: Intelligent process control using neural fuzzy techniques. J. Process Control **9**, 493–503 (1999)
33. Tchobanoglous, G., Burton, F., Stensel, H.: Wastewater Engineering: Treatment and Reuse, pp. 526–528. Metcalf & Eddy Inc., New York (2003)

34. Cărbureanu, M.: Researches on the usage of artificial intelligence techniques in wastewater treatment processes control. Doctoral Research Report, Petroleum-Gas University of Ploiesti (2012)
35. The NTPA-001/2002 normative for establishing the loading limits with pollutants for industrial and city wastewater at the evacuation into the natural receivers, published in the Romania Official, part I, no. 187, March 20, 2002, Cited 10 May 2012
36. Ibrahim, R.: Practical modeling and control implementation studies on a pH neutralization process pilot plant. Ph.D. thesis, Department Electronics and Electrical Engineering, University of Glasgow (2008)
37. Neuro-Fuzzy Systems (2012). http://www.ac.tuiasi.ro/ro/library/cursDIAGNOZAweb/p3_cap2_web.pdf, Cited 15 May 2012

Chapter 8
Application of Swarm Intelligence in Fuzzy Entropy Based Image Segmentation

R. Krishna Priya, C. Thangaraj, C. Kesavadas and S. Kannan

Abstract An image segmentation technique based on Modified Particle Swarm optimised—fuzzy entropy is applied for Infra Red (IR) images to detect the object of interest and Magnetic Resonance (MR) brain images to detect a brain tumour is presented in this chapter. Adaptive thresholding of input IR images and MR images are performed based on the proposed method. The input image is classified into dark and bright parts with Membership Functions (MF), whose member functions of the fuzzy region are Z-function and S-function. The optimal combination of parameters of these fuzzy MFs are obtained using Modified Particle Swarm Optimization (MPSO) algorithm. The objective function for obtaining the optimal fuzzy MF parameters is considered to be the maximum the fuzzy entropy. Through numerous examples, the performance of the proposed method is

R. Krishna Priya (✉)
Department of Computer Science, Kalasalingam University, Anand Nagar, Krishnankoil, Srivilliputhur, Virudhunagar 626126, Tamil Nadu, India
e-mail: priyasanchez@yahoo.com

R. Krishna Priya
Department of Applied Electronics and Instrumentation,
SAINTGITS College of Engineering, Kottayam 686532, Kerala, India

C. Thangaraj
Anna University of Technology, Chennai, India
e-mail: thangaraj.vc@gmail.com

C. Thangaraj
25/71, P.S.K Nagar, Rajapalayam 626117, Tamil Nadu, India

C. Kesavadas
Department of Imaging Sciences and Interventional Radiology, Sree Chitra Tirunal Institute for Medical Sciences and Technology, Trivandrum, Kerala, India
e-mail: chandkesav@yahoo.com

S. Kannan
Department of Electrical and Electronics Engineering, Kalasalingam University, Anand Nagar, Krishnankoil, Srivilliputhur, Virudhunagar 626126 Tamil Nadu, India
e-mail: kannaneeps@gmail.com

V. E. Balas et al. (eds.), *Innovations in Intelligent Machines-5*,
Studies in Computational Intelligence 561, DOI: 10.1007/978-3-662-43370-6_8,
© Springer-Verlag Berlin Heidelberg 2014

227

compared with those using existing entropy-based object segmentation approaches and the superiority of the proposed method is demonstrated. The experimental results obtained are compared with the enumerative search method and Otsu segmentation technique. The result shows the proposed fuzzy entropy based segmentation method optimized using MPSO achieves maximum entropy with proper segmentation of region of interest for IR images and infected areas for MR brain images with least computational time.

8.1 Introduction

The major significant and highly convoluted low-level image analysis tasks are image segmentation. The image segmentation is a process to extract meaningful objects or specified regions from an image based on threshold levels. A review by [1–3] shows threshold based segmentation is most effective. To extract a particular object from the background of an image by applying different threshold values remains as a challenge. In [4, 5] maximum entropy is derived from the histogram of an image. Among all the thresholding methods, entropy-based technique is broadly studied and is considered effective. The entropic correlation developed by Yen et al. [6] obtains an optimum threshold that maximizes it. The research works reported in [7, 8] inferred the maximization of the entropies computed from auto correlation functions to locate a threshold value to characterize the segmentation of the image. A significant role played by fuzzy sets in organizing systems with their competence to model non-statistical imprecision is presented in [9]. Luca and Termini [10] introduced the concept of fuzzy entropy. The fuzzy entropy defined on a function of fuzzy sets converges to low value when the sharpness of its fuzzy set argument is improved. There are various applications of fuzzy entropies in image segmentation. A thresholding approach based on the fuzzy relation and the maximum fuzzy entropy principle using fuzzy partition on a two-dimensional histogram has been proposed by Cheng et al. [11]. An optimum threshold is set among the least sum of entropies for an image and the importance of fuzzy memberships in indicating depth of gray value in an image's background is well expressed in [12]. The measure of compatibility among the probability partition and fuzzy c-partition was discussed by Fu [13]. A novel approach to segment MR brain image based on fuzzy entropy through probability analysis, using fuzzy partition and fuzzy entropy theory is presented. The image is partitioned into two parts, namely the dark region and the gray, where Z-MF corresponds to dark and S MF corresponds to bright region.

In this chapter, we observe the performance of segmentation techniques applied on MR brain images using Otsu method [14] and exhaustive fuzzy entropy object segmentation method. To find the optimal threshold value, it is required to search for all the possible fuzzy combinations. Thus, the segmentation problem is formulated as an optimization problem. Various researches proved that particle swarm optimization (PSO) can deploy good result for many engineering problems [15–17]. Hence a modified particle swarm optimization (MPSO) method is found to obtain effective optimal fuzzy membership parameters.

This chapter explores how MPSO is applied to find the optimal fuzzy MFs parameters to obtain maximum fuzzy entropy for the IR test images and MR brain images. This chapter is structured as follows. In Sect. 8.2, for the integrity of this chapter, we simply describe the object segmentation method based on probability analysis and fuzzy entropy, which is similar to the method presented in [18, 19]. In Sect. 8.3, the use of modified MPSO approach to find the optimal combination of all fuzzy parameters is presented. In Sect. 8.4, the performance of the proposed thresholding approach using IR test images and MR brain images is evaluated and is compared with leading techniques from the literature. Ultimately, Sect. 8.5 concludes this chapter.

8.2 Background

8.2.1 Image as a Fuzzy Event

Let A be an image of size $M \times N$ with L gray levels ranging from $L_{\min}$ to $L_{\max}$, where a_{ij} denote the gray level of the image A at the (i, j)th pixel. The histogram of the image is denoted as h_k and is defined as

$$h_k = \frac{n_k}{M \times N}, k = 0, .., L - 1 \tag{8.1}$$

where, n_k denotes the number of occurrences of gray levels in A. An image can modeled by a triplet(G, K, P), where, $G = \{r_0, r_1, r_2, ..., r_{L-1}\}$, P is the probability measure of the occurrence of gray levels, i.e. $\Pr\{r_k\} = h_k$.

For an image, a probability space based fuzzy event can be modeled. The fuzzy set theory states that an image A can be transformed into an array of fuzzy singletons S by a membership function.

$$S = \{\mu_A(a_{ij}), i = 1, 2, ..., M; j = 1, 2, ..., N\} \tag{8.2}$$

Then, the degree of some properties of the image such as brightness, darkness, etc. possessed by the (i, j)th pixel is denoted by the membership function $\mu_A(I_{ij})$ of the fuzzy set, $A \in G$. In fuzzy set notation, A can be written as

$$A = \frac{\mu_A(r_1)}{r_1} + \frac{\mu_A(r_2)}{r_2} + \cdots + \frac{\mu_A(r_k)}{r_k} \tag{8.3}$$

(Or)

$$A = \sum_{r_k} \in G \frac{\mu_A(r_k)}{r_k} \tag{8.4}$$

Here "+" indicates union.

The Equation to obtain the probability of A is given as

$$\sum_{i=0}^{L-1} \mu_A(r_k) \Pr(r_k) \tag{8.5}$$

and the Equation corresponding to conditional probability tends to be,

$$p[\{r_k\}|A] = \mu_A(r_k)h_k \big/ P(A) \tag{8.6}$$

8.2.2 Probability Partition Based Maximum Fuzzy Entropy

Fuzzy set is an extension of classical set in which Fuzzy entropy describes the fuzziness of a fuzzy set. It is a measure of the uncertainty of a fuzzy set. The domain of the image be given as Z

$$Z = \{(i,j) : i = 0, 1, 2, \ldots, M-1; j = 0, 1, 2, \ldots, N-1\} \tag{8.7}$$

and the gray level of the image as $G = \{0, 1, \ldots, L-1\}$ where M, N and L are three positive integers. If the gray level value of the image at the pixel (x, y) is $A(x, y)$ then

$$Z_k = \{(x, y) : A(x, y) = k, (x, y) \in G\}, k = 0, 1, \ldots, L-1 \tag{8.8}$$

Let T be the threshold of the image A that segments an image into its target and background. The domain Z of the original image can be classified into two parts, F_d and F_b, which is composed of pixels with low gray levels and high gray levels, respectively. An unknown probabilistic partition of Z denoted as $\prod_2 \{F_d, F_b\}$ describes its probability distribution as

$$p_d = P(F_d) \tag{8.9}$$

$$p_b = P(F_b) \tag{8.10}$$

An image with 256 gray levels is partitioned into fuzzy membership functions μ_b corresponds to bright pixels and μ_d corresponds to dark pixels. Let a, b and c be the three parameters of the membership function, which means that the threshold T depends on a, b and c. Consider

$$Z_{kd} = \{(x,y) : I(x,y) \leq T, (x,y) \in Z_k\} \tag{8.11}$$

$$Z_{kb} = \{(x,y) : I(x,y) > T, (x,y) \in Z_k\} \tag{8.12}$$

for each $k = 0, 1,..., 255$.

Then the following Equations hold:

$$p_{kd} = P(Z_{kd}) = p_k * p_{d|k} \tag{8.13}$$

$$p_{kb} = P(Z_{kb}) = p_k * p_{b|k} \tag{8.14}$$

The conditional probability of a pixel, obviously set as $p_{d|k}$ and $p_{b|k}$ is categorised into the class 'dark' and class 'bright', with the constraint that the pixel belongs to D_k with

$$p_{d|k} + p_{b|k} = 1, (k = 0, 1, \ldots, 255) \tag{8.15}$$

The grade of pixels classified into class 'dark' and class 'bright' having the gray level value k, be equal to its conditional probability $p_{d|k}$, $p_{b|k}$, respectively [3, 11, 12, 14]. The Equations for probability p_d and p_b hold as follows:

$$p_d = \sum_{k=0}^{255} p_k * p_{d|k} = \sum_{k=0}^{255} p_k * \mu_d(k) \tag{8.16}$$

$$p_b = \sum_{k=0}^{255} p_k * p_{b|k} = \sum_{k=0}^{255} p_k * \mu_b(k) \tag{8.17}$$

8.2.3 Fuzzy Membership Functions for Dark and Bright Classes

The two membership functions, S MF and Z MF are applied for calculating the fuzzy entropy function which is shown in Fig. 8.1. Here $Z(k, a, b, c)$.—function denotes the membership function $\mu_d(k)$ of the class 'dark' and $S(k, a, b, c)$—function denotes the membership function $\mu_b(k)$ of the class 'bright'. The fuzzy parameters a, b, c satisfy the constraint of an image as $0 \leq a \leq b \leq c \leq 255$. The $Z(k, a, b, c)$ membership function characteristics are given in Eq. (8.18) and the $S(k, a, b, c)$ membership function characteristics are given in Eq. (8.19).

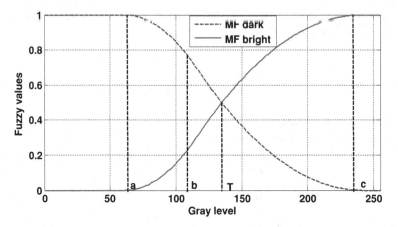

Fig. 8.1 Membership function graph showing the intersection of Z membership function- $\mu_d(k)$ and S membership function- $\mu_b(k)$ at T

$$\mu_d(k) = \begin{cases} 1, & k \leq a \\ 1 - \dfrac{(k-a)^2}{(c-a)*(b-a)}, & a < k \leq b \\ \dfrac{(k-c)^2}{(c-a)*(c-b)}, & b < k \leq c \\ 0, & k > c \end{cases} \tag{8.18}$$

$$\mu_b(k) = \begin{cases} 0, & k \leq a \\ \dfrac{(k-a)^2}{(c-a)*(b-a)}, & a < k \leq b \\ 1 - \dfrac{(k-c)^2}{(c-a)*(c-b)}, & b < k \leq c \\ 1, & k > c \end{cases} \tag{8.19}$$

Fuzzy entropy function for dark class H_d and bright class, H_b calculated based on Eqs. (8.20) and (8.21) is given below as:

$$H_d = -\sum_{k=0}^{255} \frac{p_k * \mu_d(k)}{p_d} * \log\left(\frac{p_k * \mu_d(k)}{p_d}\right) \tag{8.20}$$

$$H_b = -\sum_{k=0}^{255} \frac{p_k * \mu_b(k)}{pb} * \log\left(\frac{p_k * \mu_b(k)}{p_b}\right) \tag{8.21}$$

Then the total fuzzy entropy function $H(a, b, c)$ is given as

$$H(a,b,c) = H_d + H_b \qquad (8.22)$$

The obtained total fuzzy entropy depends on the fuzzy parameters—a, b, c. The combination of these three parameters is chosen at the point where the fuzzy entropy $H(a, b, c)$ attains a maximum value. The Equation to segment the image into two classes using appropriate threshold is as follows:

$$\mu_d(T) = \mu_b(T) = 0.5 \qquad (8.23)$$

It is evident from the Fig. 8.1, that threshold T is the point of intersection of $\mu_d(k)$ and $\mu_b(k)$ curve. The solution to derive T can be obtained from the Eq. (8.24). Hence the threshold T can be easily formulated by:

$$T = \begin{cases} a + \sqrt{(c-a)*(b-a)/2}, & (a+c)/2 \leq b \leq c \\ c - \sqrt{(c-a)*(c-b)/2}, & a \leq b \leq (a+c)/2 \end{cases} \qquad (8.24)$$

8.2.4 Modified Particle Swarm Optimization Algorithm

PSO developed by Eberhart and Kennedy [15] is a population based, stochastic search technique inspired by social behaviours of animals such as bird flocking and fish schooling. It is similar to other population based optimization methods, PSO starts with the random initialization of a population in the search space. Based on the social behaviour of particles in the swarm a PSO algorithm is framed. The most prominent of these are its characteristics towards stable convergence to generate a high quality solution in a shorter execution time than other stochastic methods.

The modified concept of search point by PSO is shown in Fig. 8.2.

Where $_nx_i^d$ is the current position, $_{n+1}x_i^d$ is modified position, $_nv_i^d$ is the current velocity, $_{n+1}v_i^d$ is the modified velocity, v_i^{pbest} is the velocity based on $pbest_i$ and v_i^{gbest} is the velocity based on $gbest_d$.

The velocity v_i^d and positions x_i^d are up dated based on the Equation given below, [15]:

$$_{n+1}v_i^d = \omega_n v_i^d + c_1 r_{1_i}^d (pbest_i^d - _nx_i^d) + c_2 r_{2_i}^d (gbest^d - _nx_i^d) \qquad (8.25)$$

$$_{n+1}x_i^d = _nx_i^d + _{n+1}v_i^d \qquad (8.26)$$

$$i = 1, 2, 3 \dots, N, \quad d = 1, 2, 3 \dots, D.$$

Fig. 8.2 Concept of
modification of a search point
by PSO

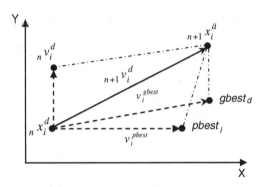

Where $x_i = \left(x_i^1, x_i^2, x_i^3, \ldots, x_i^D\right)$ is the position of the ith particle, $pbest_i = (pbest^1, pbest^2, \ldots, pbest^D)$ is the best local best position of a particle, $gbest = (gbest^1, gbest^2, \ldots, gbest^D)$ is the global best position discovered by the entire population, $v_i = \left(v_i^1, v_i^2, v_i^3, \ldots, v_i^D\right)$ is the velocity of a particle i, c_1 and c_2 are the acceleration constants, n is the migration number, r_1 and r_2 are the random variables and ω is the inertia weight.

In modifying the standard PSO, a linearly time-varying acceleration constant is introduced in evolutionary procedure as suggested in [20, 21] is applied, hence MPSO. The MPSO modifies the Eq. (8.25) for a high cognitive constant (c_1) and low social constant (c_2) at the start of the algorithm, and gradually c_1 is decreased and c_2 is increased to move the particle around the entire search space instead of converging towards a local minima. Finally in the optimization, the particles are allowed converge to the global optima.

$$c_1(iter) = (c_{1,min} - c_{2,max})\frac{iter}{iter_{max}} + c_{1,max} \qquad (8.27)$$

$$c_2(iter) = (c_{2,max} - c_{1,min})\frac{iter}{iter_{max}} + c_{2,min} \qquad (8.28)$$

where $iter$ is the current iteration number and $iter_{max}$ is the maximum iteration number.

Then $_{n+1}v_i^d$ and $_{n+1}x_i^d$ should be under the constrained conditions as follows:

$$_{n+1}v_i^d = \begin{cases} _{n+1}v_i^d, & -v_{max} \leq\ _{n+1}v_i^d \leq v_{max} \\ v_{max}, & _{n+1}v_i^d > v_{max} \\ -v_{max}, & _{n+1}v_i^d < -v_{max} \end{cases} \qquad (8.29)$$

$$_{n+1}x_i^d = \begin{cases} _{n+1}x_i^d, & x_{min} \leq\ _{n+1}x_i^d \leq x_{max} \\ x_{init}, & _{n+1}x_i^d > x_{max} \\ x_{init}, & _{n+1}x_i^d < x_{min} \end{cases} \qquad (8.30)$$

$$x_{init} = x_{min} + rand() * (x_{max} - x_{min}) \tag{8.31}$$

where v_{max} is the maximum value of v; x_{max} and x_{min} are the maximum and minimum value of x, respectively.

8.3 Methodology

8.3.1 Fuzzy Parameter Optimization Using MPSO

The three fuzzy parameters (a, b and c) are used to design fuzzy MFs. The two membership functions Z MF and S MF are constructed by these three parameters subject to the constraint $0 \leq a < b < c \leq 255$. These three parameters are optimized using MPSO. The MPSO obtains optimal solution using fuzzy entropy as the objective function based on Eq. (8.22). This optimization is considered as a minimization problem hence the fitness function is considered as inverse of objective function. The threshold is calculated from the optimal fuzzy MFs parameters and segmentation of region of interests of both IR test images and MR brain images are carried out. The process for obtaining the optimal threshold based on maximum entropy using MPSO is illustrated in Fig. 8.3.

The procedure can be summarized as follows:

Initialization of the particle swarm for the position matrix X and the velocity matrix V are given in the equations below as:

$$_nx_i^d = x_{min} + (x_{max} - x_{min}) * rand() \tag{8.32}$$

$$X = \begin{bmatrix} x_{11} & x_{12} & x_{13} \\ x_{21} & x_{22} & x_{23} \\ \cdots & \cdots & \cdots \\ x_{N1} & x_{N2} & x_{N3} \end{bmatrix} \tag{8.33}$$

$$_nv_i^d = -v_{max} + 2v_{max} * rand() \tag{8.34}$$

$$V = \begin{bmatrix} v_{11} & v_{12} & v_{13} \\ v_{21} & v_{22} & v_{23} \\ \cdots & \cdots & \cdots \\ v_{N1} & v_{N2} & v_{N3} \end{bmatrix} \tag{8.35}$$

where, x_{max} and x_{min} are the maximum and minimum value of position (x) where $x_{max} = L_{max}, x_{min} = L_{min} + 1$, $x_{i2} - x_{i1} \geq 2$ and $x_{i3} - x_{i2} \geq 2$; L_{max} and L_{min} are the corresponding maximum and minimum gray levels of the image. Fitness value is calculated for each particle using the fuzzy entropy function. The comparison is made among the evaluated current fitness values with that of the fitness value of its best previous position. If the current fitness value is found to be better, then the best

Fig. 8.3 MPSO based fuzzy
entropy image segmentation
overview

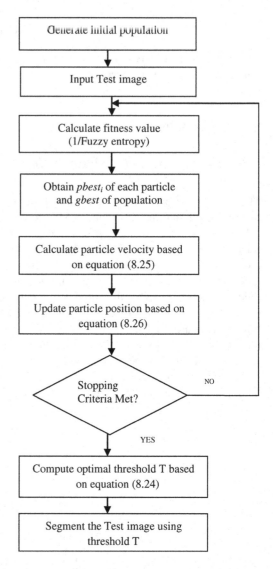

previous position is set as the current best position. Then compare the evaluated
fitness value of each particle with the fitness value of the whole swarm's best previous
position, *pbest*. If the current value is better, subsequently set the current position as
the whole swarm's best previous position. Subject to the constraints of Eqs. (8.29)
and (8.30), an updating of the velocity of each particle is carried out using to
Eq. (8.25) and updating of the position of each particle is carried out with Eq. (8.26).
The predefined maximum iterative time is the stopping criterion. If the terminating
criterion is not satisfied, then the MPSO will search for the next best particle in the
swarm. When the terminating criterion is satisfied, the threshold T is calculated based
on the optimal fuzzy MF parameters (a, b, c) then segmentation is carried out.

8.4 Experimental Results

The entire simulation is carried out using MATLAB 7.10 on a Desktop with Intel®
Core™ i5-Processor and 4 GB RAM. The MPSO parameters were initialized as
mentioned in Table 8.1.

In substantiating this work, a set of IR test images and MR brain images is used
as the experimental data. The IR test images include a region of interests to be
segmented out from the back ground and diseased parts that should be segmented
from the brain in the MR brain test images. In order to authenticate the effectiveness of the proposed method, the performance of segmentation is compared
with that of existing methods such as, exhaustive search method and Otsu's segmentation method [14]. The simulation results of IR test images and MR brain
images are shown in Figs. 8.4, 8.5, 8.6 and 8.7. Each figure shows the test image,
the segmented images using the proposed method as well as other methods used in
comparison. The test images used for segmentation include IR images of a tank,
Fig. 8.4a and person image, Fig. 8.5a, then MR brain images of tumour in the left
medial parietal cortex, shown in Fig. 8.6a and the tumour located in the supersellar
region in Fig. 8.7a. These images were taken from online available IR images data
base and MNI BrainWeb.

Figure 8.4a shows the IR test image 1 of Tank. Figure 8.4b shows the segmentation performed by the proposed method with an obtained optimal threshold
value of T = 144.7 and Fig. 8.4c shows the segmentation performed by the
exhaustive search method with a threshold value of T = 144.7. Figure 8.4d shows
the segmentation performed by Otsu segmentation method with a threshold value
of T = 44.12. From the output of segmentation revealed in the Fig. 8.4b, c and d,
it is inferred that the region of interest area is more precisely segmented using the
proposed method, whereas the performances of Otsu segmentation method was not
acceptable. The comparison of the results for the IR test image 1 is depicted in
Table 8.2.

The comparison analysis of IR test image 1 is shown in Table 8.2 using the
three methods with respect to threshold, entropy and computational time. Though
Otsu segmentation method requires a very low computational time to segment, the
results are not acceptable. Whereas the computational time required by MPSO-
fuzzy entropy method is comparably less when compared that of other two
methods. The exhaustive search method is one of the well proven conventional
search methods to find the optimal value in entire search space. The exhaustive
search method gives maximum entropy of 9.601; however the computational time
required in finding the best values is very high. The proposed MPSO method gives
the same threshold as that of exhaustive search method with minimum computational time which is 125 times lesser than that of exhaustive search method.

Figure 8.5a shows the IR test image 2 (Person). Figure 8.5b shows the segmentation performed by the proposed method with an obtained optimal threshold
value of T = 99.29 and Fig. 8.5c shows the segmentation performed by the
exhaustive search method with a threshold value of T = 99.29. Figure 8.5d shows

Table 8.1 MPSO Parameter settings

Parameters		Value
Swarm size		25
Self-recognition coefficient, c_1	$c_{1\ min}$	0.5
	$c_{1\ max}$	2.5
Social coefficient, c_2	$c_{2\ min}$	0.5
	$c_{2\ max}$	2.5
Inertia weight, ω		1
Bird step		150

(a)   **(b)**   **(c)**   **(d)**

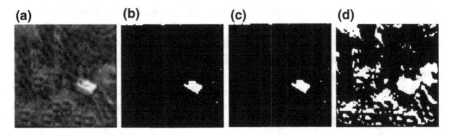

Fig. 8.4 **a** IR Test image 1 Tank (225 × 225), **b** thresholding result by the proposed method (T = 144.7), **c** results by exhaustive search method (T = 144.7), **d** Otsu segmentation (T = 44.12)

(a)   **(b)**   **(c)**   **(d)**

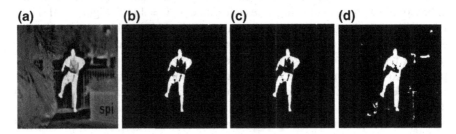

Fig. 8.5 **a** IR Test image 2 Person (320 × 210), **b** thresholding result by the proposed method (T = 99.29), **c** results by exhaustive search method (T = 99.29), **d** Otsu segmentation (T = 21.32)

(a)   **(b)**   **(c)**   **(d)**

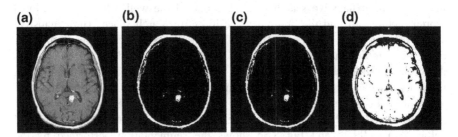

Fig. 8.6 **a** MR brain Test image 1 (225 × 225), **b** thresholding result by the proposed method (T = 184.142), **c** results by exhaustive search method (T = 184.142), **d** Otsu segmentation (T = 74.2902)

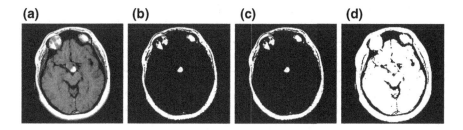

Fig. 8.7 **a** MR brain Test image 2 (630 × 612 × 3), **b** thresholding result by the proposed method (T = 193.7421), **c** results by exhaustive search method (T = 193.7421), **d** Otsu segmentation (T = 82.3216)

Table 8.2 Comparison of results for the IR test image 1 Tank

IR Image 1 (148 × 160)				
Methodology	Fuzzy MF parameters (a, b, c)	Threshold (T)	Entropy (H)	Time (s)
Otsu segmentation	NA	44.12	2.89	0.4194
Exhaustive search	(159, 247, 251)	144.7	9.601	338.211
MPSO	(159, 247, 251)	144.7	9.601	3.563

Table 8.3 Comparison of results for the IR test image 2

IR Image 2 (320 × 210)				
Methodology	Fuzzy MF parameters (a, b, c)	Threshold (T)	Entropy (H)	Time (s)
Otsu Segmentation	NA	21.32	5.6704	0.4281
Exhaustive search	(69, 71, 170)	99.29	9.3224	293.622
MPSO	(69, 71, 170)	99.29	9.3224	3.9421

the segmentation performed by Otsu segmentation method with a threshold value of T = 21.32. By observing Fig. 8.5b, c and d, the region of interests of IR test image 2 is more precisely segmented using the proposed method. The comparison of the results for the IR test image 2 (Person) is depicted in Table 8.3.

Table 8.3 shows the comparison of the results obtained using the three methods with respect to threshold, entropy and computational time for the IR test image 2. The performance of the proposed method gives the improved results when compared with other two methods. The proposed method obtains the equivalent threshold as that of exhaustive search method with minimum computational time which is 74 times lesser than that of exhaustive search method.

Figure 8.6a shows the MR brain test image 1. Figure 8.6b shows the segmentation performed by the proposed method with an obtained optimal threshold value of T = 184.142 and Fig. 8.6c shows the segmentation performed by the

Table 8.4 Comparison of results for MR brain test image 1

MR Image 1 (225 × 225)				
Methodology	Fuzzy MF parameters (a, b, c)	Threshold (T)	Entropy (H)	Time (s)
Otsu segmentation	NA	74.2902	4.8774	0.3273
Exhaustive search	(99, 203, 238)	184.142	7.4158	353.6871
MPSO	(99, 203, 238)	184.142	7.4158	2.0842

Table 8.5 Comparison of results for MR brain test image 2

MR Image 2 (630 × 612 × 3)				
Methodology	Fuzzy MF parameters (a, b, c)	Threshold (T)	Entropy (H)	Time (s)
Otsu segmentation	NA	82.3216	4.9346	0.5149
Exhaustive search	(95, 225, 245)	193.7421	6.9207	358.0983
MPSO	(95, 225, 245)	193.7421	6.9207	4.7685

exhaustive search method with a threshold value of T = 184.142. Figure 8.6d shows the segmentation performed by Otsu segmentation method with a threshold value of T = 74.2902. By observing these figures, the Fig. 8.6b and c shows the segmentation of the infected area more precisely, whereas the performances of Otsu segmentation method was not acceptable. The comparison of the results for MR brain test image 1 is depicted in Table 8.4.

Table 8.4 shows the comparison of the results obtained using the three methods with respect to threshold, entropy and computational time for the MR brain test image 1. The comparison results prove that the proposed method outperformed the Otsu segmentation method with maximum entropy and with minimum computational time. The proposed MPSO method gives the same threshold as that of exhaustive search method with minimum computational time which is 170 times lesser than that of exhaustive search method.

Figure 8.7a shows the MR brain test image 2. Figure 8.7b shows the segmentation performed by the proposed method with an obtained optimal threshold value of T = 193.7421 and Fig. 8.7c shows the segmentation performed by the exhaustive search method with a threshold value of T = 193.7421. Figure 8.7d shows the segmentation performed by Otsu segmentation method with a threshold value of T = 82.3216. It is observed from Fig. 8.7b, c and d that the segmentation of the infected area was performed more precisely in the proposed method. The comparison of the results for the MR brain test image 2 is depicted in Table 8.5.

Table 8.5 shows the comparison of the results obtained using the three methods with respect to threshold, entropy and computational time for the MR brain test image 2. The performance of the proposed method gives the better results when

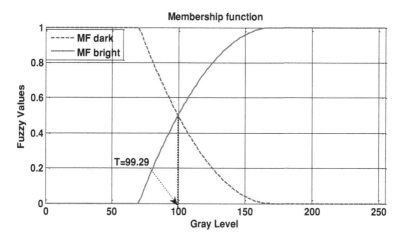

Fig. 8.8 The membership curves of the IR test image 1 with a = 69, b = 71, c = 170 and maximal fuzzy entropy H = 9.3224

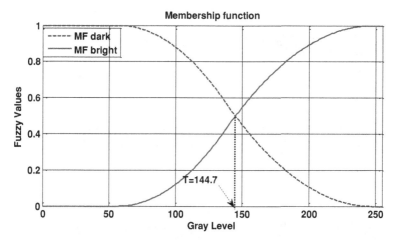

Fig. 8.9 The membership curves of the IR test image 2 with a = 57, b = 137, c = 247 and maximal fuzzy entropy H = 9.6012

compared with other two methods. The proposed method gives the same threshold as that of exhaustive search method with minimum computational time which is 75 times lesser than that of exhaustive search method.

The optimal fuzzy membership classification for the IR and MR test images are shown in following Figs. 8.8, 8.9, 8.10 and 8.11. These fuzzy membership function classification yields the maximum entropy for the corresponding test images.

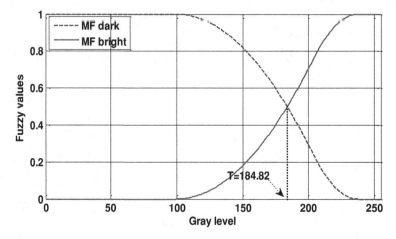

Fig. 8.10 The membership curves of the image 1 with a = 99, b = 203, c = 238 and maximal fuzzy entropy H = 7.4158

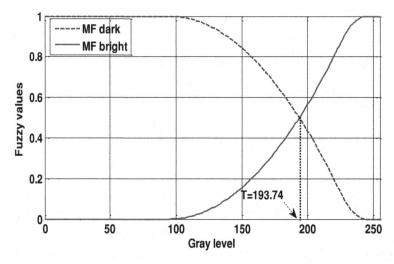

Fig. 8.11 The membership curves of the image 2 with a = 95, b = 225, c = 245 and maximal fuzzy entropy H = 6.9207

8.4.1 Convergence Test

In order to validate the randomness of the proposed methodology, a frequency of convergence to the near optimal solutions for the test images with 25 trial runs were performed according to the parameter settings given in Table 8.6. The convergence results are summarized in Table 8.6.

Table 8.6 Convergence results with test images for 25 trial runs

| Test Image | Fuzzy membership function parameters | | | | | | | | | | Fuzzy entropy (H) | | |
| | a | | | | b | | | | c | | | | |
	50–120	121–191	192–262	70–140	141–211	212–282	170–240	226–240	241–311		6.9–7.8	7.9–8.8	8.9–9.8
IR Image 1	22	3	0	21	2	2	21	3	1		0	2	23
IR Image 2	23	2	0	22	3	0	23	2	0		0	1	24
MR Image 1	23	2	0	1	2	22	22	2	1		23	2	0
MR Image 2	24	1	0	0	1	24	21	3	1		22	2	1

8.5 Conclusion

In this chapter, a bi-level thresholding method for IR images and MR brain images segmentation based on maximum fuzzy entropy was introduced. A modified PSO was introduced to have effective exploration and exploitation. The optimal fuzzy membership function parameters are optimized using MPSO. To analyse the performances of the proposed method, the results were compared with that of conventional adaptive thresholding method. The results show the proposed method obtains satisfactory performances in the segmentation experiments conducted for different test images. In order to ensure the optimized fuzzy parameters are global optimum, the results are compared with conventional search method (enumerative search method). The proposed method is competent enough in finding the global optimal fuzzy membership parameters as that of the conventional search method with minimum computational time. Due to the randomness in validating the consistency and robustness of the proposed method, convergence test are carried out. Inference from the results shows that more than 95 % of the results are identical as the conventional method. Hence, it is concluded that fuzzy entropy integrated MPSO based image segmentation technique is one of the effective method for bi-level segmentation and can be adapted for segmenting the region of interests in IR images and to segment the infected portion of the MR brain images efficiently. This method can also be extended to segment other sort of images such satellite images, other medical images etc. Thus, this method proves to be an alternate to bi-level segmentation to acquire improved performances in any image processing applications.

References

1. Sahoo, P.K., Solutani, S., Wong, A.K.C.: A survey of thresholding techniques. Comput. Vis. Graph. Image Process. **41**, 233–260 (1988)
2. Pal, N.R., Pal, S.K.: A review on image segmentation techniques. Pattern Recogn. **26**, 1277–1294 (1993)
3. Sezgin, M., Sankur, B.: Survey over image thresholding techniques and quantitative performance evaluation. J. Electron. Imaging **13**(1), 146–165 (2004)
4. Pun, T.: A new method for gray-level picture threshold using the entropy of the histogram, Signal. Process. **2**(3), 223–237 (1980)
5. Pun, T.: Entropic thresholding: a new approach, Comput. Graph. Image Process. **16**, 210–239 (1981)
6. Chang, F.J., Chang, S., Yen, J.C.:A new criterion for automatic multilevel thresholding. IEEE Trans. Image Process. IP-**4**, 370–378 (1995)
7. Sahoo, P., Wilkins, C., Yeager, J.: Threshold selection using renyi's entropy. Pattern Recogn. **30**, 71–84 (1997)
8. Brink, A.D., Pendock, N.E.: Minimum cross entropy threshold selection. Pattern Recogn. **29**, 179–188 (1996)
9. Ebanks, B.R.: On measures of fuzziness and their representations. J. Math. Anal. Appl. **94**, 24–37 (1983)

10. Luca, A.D., Termini, S.: Definition of a non probabilistic entropy in the setting of fuzzy sets theory. Inf. Contr. **20**, 301–315 (1972)
11. Cheng, H.D., Chen, Y.H., Jiang, X.H.: Thresholding using two dimensional histogram and fuzzy entropy principle. IEEE Trans. Image Process. **9**(4), 732–735 (2000)
12. Jayaraman, V.K., Kulkarni, B.D., Shelokar, P.S.: An ant colony approach for clustering. Anal. Chim. Acta **59**, 187–195 (2004)
13. Fu, A., Yan, H., Zhao, M.: A technique of three-level thresholding based on probability partition and fuzzy 3-partition, IEEE Trans. Fuzzy Syst. **9**(3), 469–479 (2001)
14. Otsu N.: A threshold selection method from gray-level histograms. IEEE Trans. Syst. Man Cybern.**9**(1), 62–66 (1979)
15. Eberhart, R.C., Kennedy, J.: A new optimizer using particle swarm theory. In: Proceedings 6th International Symposium Micromachine Human Science, pp. 39–43. Nagoya, Japan. (1995)
16. Eberhart, R.C., Kennedy, J., Shi, Y.H.: Swarm Intelligence. Morgan Kaufmann, San Mateo (2001)
17. Eberhart R.C., Shi Y.H.: Particle swarm optimization: developments, applications and resources. In: Proceedings of the IEEE Congress Evolutionary Computation,Seoul, pp. 81–86. Korea. (2001)
18. Tao, W.B., Tian, J.W., Liu, J.: Image segmentation by three-level thresholding based on maximum fuzzy entropy and genetic algorithm. Pattern Recogn. Lett. **24** (16), 3069–3078 (2003)
19. Halgamug, S. K., Ratnaweera, A., Watson. C.:Self-organizing hierarchical particle swarm optimizer with time-varying acceleration coefficients. IEEE Trans. Evol. Comput. **8**(3), 240–255 (2004)
20. Tao, W.B., Tian, J.W., Liu, J.: Object segmentation using ant colony optimization algorithm and fuzzy entropy. Pattern Recogn. Lett. **28**, 788–796 (2007)
21. Godwin Anand, P.S., Subbaraj, P.: Evolutionary design of IFLC for a three tank system. IJCSI Int. J. Comput. Sci. Issues Spec. Issue, ICVCI-2011 **1**(1) (2011)

About the Editors

Valentina E. Balas is currently a Full Professor in the Department of Automatics and Applied Software at the Faculty of Engineering, University "Aurel Vlaicu" Arad (Romania).

She holds a Ph.D. in Applied Electronics and Telecommunications from Polytechnic University of Timisoara. She is author of more than 160 research papers in refereed journals and International Conferences. Her research interests are in Intelligent Systems, Fuzzy Control, Soft Computing, Smart Sensors, Information Fusion, Modeling and Simulation.

She is the Editor-in Chief to *International Journal of Advanced Intelligence Paradigms* (*IJAIP*), member in Editorial Board member of several national and international journals and is evaluator expert for national and international projects.

Dr. Balas participated in many international conferences as General Chair, Organizer, Session Chair and member in International Program Committee.

She was a mentor for many student teams in Microsoft (Imagine Cup), Google and IEEE competitions in the last years.

She is a member of EUSFLAT, ACM and a Senior Member IEEE, member in TC—Fuzzy Systems (IEEE CIS), member in TC—Emergent Technologies (IEEE CIS), member in TC—Soft Computing (IEEE SMCS) and also a member in IFAC—TC 3.2 Computational Intelligence in Control.

Dr. Balas is Vice-president (Awards) of IFSA International Fuzzy Systems Association Council and Joint Secretary of the Governing Council of Forum for Interdisciplinary Mathematics (FIM),—A Multidisciplinary Academic Body, India.

V. E. Balas et al. (eds.), *Innovations in Intelligent Machines-5*,
Studies in Computational Intelligence 561, DOI: 10.1007/978-3-662-43370-6,
© Springer-Verlag Berlin Heidelberg 2014

Petia Koprinkova-Hristova received MSc degree in Biotechnics from the Technical University—Sofia in 1989 and PhD degree on Process Automation from Bulgarian Academy of Sciences in 2001. Since 2003 she is Associate Professor in the Institute of Control and System Research and from January 2012—in the Institute of Information and Communication Technologies, Bulgarian Academy of Sciences. Her main research interests are in the field of Intelligent Control Systems using mainly fuzzy, neuro-fuzzy and neural network approaches. Currently she is a member of European Neural Network Society (ENNS) executive committee for 2011–2016 and member of the Union of Automatics and Informatics in Bulgaria.

Lakhmi C. Jain is with the Faculty of Education, Science, Technology and Mathematics at the University of Canberra, Australia and University of South Australia, Australia. He is a Fellow of the Institution of Engineers Australia.

Dr Jain founded the KES International for providing a professional community the opportunities for publications, knowledge exchange, cooperation and teaming. Involving around 5,000 researchers drawn from universities and companies world-wide, KES facilitates international cooperation and generate synergy in teaching and research. KES regularly provides networking opportunities for professional community through one of the largest conferences of its kind in the area of KES. www.kesinternational.org

His interests focus on the artificial intelligence paradigms and their applications in complex systems, security, e-education, e-healthcare, unmanned air vehicles and intelligent agents.

Printed in the United States
By Bookmasters